Studies on Inbreeding

by Helen Dean King

with an introduction by Jackson Chambers

Self Reliance Books

Get more historic titles on animal and stock breeding, gardening and old fashioned skills by visiting us at:

http://selfreliancebooks.blogspot.com/

Introduction

I am pleased to present yet another title on the Principles of Animal Breeding.

This volume is entitled "Studies on Inbreeding" and was published in 1919.

The work is in the Public Domain and is re-printed here in accordance with Federal Laws.

As with all reprinted books of this age that are intended to perfectly reproduce the original edition, considerable pains and effort had to be undertaken to correct fading and sometimes outright damage to existing proofs of this title. At times, this task is quite monumental, requiring an almost total "rebuilding" of some pages from digital proofs of multiple copies. Despite this, imperfections still sometimes exist in the final proof and may detract from the visual appearance of the text.

I hope you enjoy reading this book as much as I enjoyed making it available to readers again.

Jackson Chambers

Kellerstrass Farm
Arthur Oscar Schilling
1907

STUDIES ON INBREEDING

I. THE EFFECTS IN INBREEDING ON THE GROWTH AND VARIABILITY IN THE BODY WEIGHT OF THE ALBINO RAT

HELEN DEAN KING

The Wistar Institute of Anatomy and Biology

The rapid development of the new science of genetics has opened up many fertile fields of investigation and it has also revived interest in the problem of inbreeding which has been dormant for many years. Charles Darwin ('75, '78) considered the subject of inbreeding so important that not only did he collect all available data regarding it, but he himself carried on a series of inbreeding experiments that extended over a period of eleven years. Darwin's experiments on plants were followed by those of Crampe ('83), of Huth ('87), and of Retzima-Bos ('93; '94) on various species of mammals. The conclusions reached by each of these investigators can well be stated in the words of Darwin ('78): "The consequences of close interbreeding carried on for too long a time, are, as is generally believed, loss of size, constitutional vigor and fertility, sometimes accompanied by a tendency to malformation." Darwin adds, furthermore: "That any evil directly follows from the closest interbreeding has been denied by many persons, but rarely by any practical breeder and never, as far as I know, by one who has largely bred animals which propagate their kind quickly."

On account of the almost universal prejudice against inbreeding, or because the results of former work seemed conclusive, the problem of inbreeding was practically ignored by scientists after the publication of Retzima-Bos' results in 1893, and only within the past decade has it again received any serious consideration. The recent experiments of Gentry ('05) on swine, of Castle et al. ('06) and of Moenkhaus ('11) on Drosophila, and brother and sister from the same litter is the closest form of inbreeding possible in mammals, no other form of mating has ever been used to obtain inbred litters. In every generation all females used for breeding have belonged to inbred litters; none of them have ever been taken from 'half-inbred' litters obtained by the mating of inbred females with stock males. In the early part of the experiment the number of breeding animals was, of necessity, small. In every generation after the sixth about twenty females from each series were used for breeding, so that approximately 1000 young were obtained in each generation.

All four of the rats used in starting the experiment were killed when they were no longer wanted for breeding purposes. Each rat was weighed, measured, and carefully dissected. When the various records were compared with the norms for the albino rat (Donaldson, '15) it was found that all of the rats were under the average body weight for their age, but that they were normal in all other respects as far as could be determined by the usual methods of laboratory procedure. The fact that these individuals were sound, healthy animals and normal in all essential respects is a point on which I wish to place special emphasis in order to forestall the possible criticism that the results obtained in this work were due to the use of an exceptional strain of rats.

In the earlier generations the inbred rats exhibited all of the defects which are popularly supposed to appear in any closely inbred stock. Many females in both series were sterile, and those that did breed usually produced only one or two litters which were generally of small size. A considerable proportion of the rats were dwarfed, or stunted in their growth, and many of them developed malformations, particularly deformed teeth. The animals showed, also, a steady decline in vitality in succeeding generations and usually died at a relatively early age. If the experiments had been discontinued at this point the results would have been a confirmation of the conclusion reached by Darwin and by several other investigators, that inbreeding invariably leads to sterility and to physical degeneration.

Fortunately for this work, many rats in the general stock colony, in which there was no inbreeding, exhibited the same characteristics as the rats belonging to the inbred strain. It was evident, therefore, that the unfavorable condition of the animals could not justly be attributed to inbreeding alone. On investigation it was found that all of the rats were suffering from malnutrition due to the character of the food that they received. At the time that these experiments were begun the rat was used as a laboratory mammal in only a few of the larger research institutions in the country, and little was known of the environmental and nutritive conditions best suited to its needs. Following the plan of feeding in general use in other animal colonies, the rats were fed chiefly on bread soaked in milk and on corn; meat and vegetables being given only once each week. Such a diet does not furnish the proportion of food elements that the rat requires if it is to be kept in good physical condition: there is too much starch, too little protein. In the spring of 1911 a radical change was made in the rats' food. Milk and fresh bread were eliminated from the diet and 'scrap' food, consisting of carefully sorted table refuse, was fed once each day; cobcorn being kept in the cages as an extra ration. Such a diet has proved to be a most satisfactory one, and it has been used continually up to the present time, except that dog biscuit has been substituted for cobcorn as extra food supply. This change was made last year following the loss of a considerable number of animals through intestinal disturbance caused by the eating of fermented corn.

A very marked improvement in the general condition of all of the rats in the colony was noted very soon after the diet was changed. The animals gained in size and in weight, sterility almost disappeared, and the average number of young in the litters was increased. From this time on malformations were no longer common, and not a single instance of deformed teeth has been discovered in the thousands of animals that have been bred in the colony during the past five years. Simply by a change in the food characteristics said to typify the 'dire effects of inbreeding' were eliminated, and up to the present time they have not reappeared, although the rats have been carried through twenty-eight generations of brother and sister matings.

In the inbred colony, up to the sixth generation, very little selection of breeding animals was possible; any females that would breed at all were used to continue the series. The change in food was made at the time that the animals of the fourth inbred generation were reaching maturity. In the course of the two following generations the effects of malnutrition gradually disappeared, and in the sixth generation most of the rats were of normal size and relatively large litters were being produced. After this time large and vigorous animals were available for breeding purposes, and it became possible to make a careful selection of the breeding stock.

From the seventh generation on the selection of the individuals which were to serve as progenitors of the succeeding generation was always made among the newborn young, as the sexes can readily be\distinguished at this time (Jackson, '12). In the A series of inbreds, which is called the 'male line,' all litters containing an excess of female young were always discarded; in the B series, the 'female line,' litters with an excess of male young were never reared. Unless the individuals in the litter were of normal size and vigorous at birth they were killed at once. The young which were retained remained with their mother until they were one month old, when they were again carefully examined, and if they did not come up to the norms for stock animals of like age they were discarded. If the young rats fulfilled all requirements as to body weight and vigor they were returned to the cage to be reared as possible breeding stock. This rigid selection left in each generation, as a rule, at least three times the number of animals that were required for breeding. When the rats become sexually mature, at about three months of age, they were again inspected, and any that were below normal in any way were rejected. Generally only one female of a litter, the first to breed, was taken to continue the line. If, however, the individuals were unusually large and vigorous, two, very rarely three, breeding females were taken from the same litter.

Since the change in the food in 1911 and the removal of the colony to new quarters in 1913, the environmental conditions under which the rats were reared have been as uniform as it was possible to make them.   All inbred rats, and also the stock animals used for controls, have been subjected to the same conditions of light, of temperature, and of nutrition, and they have been cared for in a similar way.   Any differences between the two inbred series, or between inbred and stock animals must, therefore, be ascribed to causes inherent in the individuals; they cannot be attributed to the varying action of environment or nutrition.

2. THE GROWTH IN BODY WEIGHT OF INBRED RATS

In view of the results that earlier investigators (Crampe, Ritzema-Bos) obtained in their inbreeding experiments with rats, little attention was paid to the fact that the body weights of the animals in the earlier inbred generations were considerably less than the norms for stock albino rats of like age.   When the individuals of the sixth inbred generation became mature, it was noted that many of them were much larger than stock animals of the same age.   This fact was so at variance with the generally accepted belief regarding the effects of close inbreeding on body size that it seemed desirable to make a study of the weight increase with age of individuals in the later generations of the two inbred series.

From the seventh generation on from three to five litters of each inbred series were weighed, first when the animals were thirteen days old, again when they were weaned at thirty days of age, and thereafter at intervals of one month until they were fifteen months old.   At the thirteen- and thirty-day periods animals of the same sex were weighed together and the average body weight for the group recorded, as at these ages individual differences in body weight are, as a rule, too small to make separate weighings necessary; at all other ages individual records were taken.

The litters that were used for a study of growth in body weight were all selected at birth on the same basis as the litters that were to be reared for breeding purposes. There was no culling of the less desirable individuals, however, and all members of every litter were reared and weighed at the ages noted. When the animals became mature the largest and most vigorous pair in each litter was usually used for breeding.

As the growth of very young rats depends largely on the amount of nourishment that they receive from their mother, litters of medium size, containing from five to eight young, were, as a rule, those selected for weighing. Such litters, moreover, represent the general run of individuals in a colony more fairly than do very large or very small litters in which animals with extreme body weights are often found. It was not always possible to weigh adult animals at exactly the ages designated in the various tables, but the weighing of a litter was omitted if it could not be taken within one week of the time specified.

In weighing experiments of this kind there is always an unavoidable error due to the presence of a greater or a less amount of undigested food in the alimentary tract. To obviate this source of error as far as possible the rats were weighed in the morning before they had received their daily food ration. Animals that were obviously ill and females known to be pregnant were never weighed, while the weight of suckling mothers was not recorded if it was below the previous record. In the rat pregnancy cannot be detected, with certainty until about the thirteenth day, so undoubtedly many gravid females were weighed unknowingly during the course of this investigation. The increase in the body weight of a female as the result of pregnancy cannot be very great up to the thirteenth day, however, since Stotsenburg ('15) has shown that the weight of a fetus at this time is only 0.04 grams. Errors in the records due to the inclusion of the weights of pregnant females were doubtless balanced by the weights for animals that were in early stages of pneumonia when there was no external evidence of the disease.

The present paper gives data showing the weight increase with age in 333 males and in 306 females belonging to the first fifteen generations of the inbred group. Altogether these generations comprised a total of 1601 litters, containing 11,657 individuals, of which the majority were the progeny of brother and sister matings, the others were the offspring of inbred females and stock males. The number of animals for which weight records were taken is, it is hoped, large enough to be fairly representative of the inbred colony as a whole and to give results that have some statistical value.

No animals belonging to the first six inbred generations were weighed at regular intervals, but fortunately a series of weight records is available that will give some idea of the size of these rats after they became adult. In order to ascertain the effects of close inbreeding on the weight of the central nervous system, one of my colleagues, Dr. S. Hatai, made careful autopsies of the four rats used in starting the experiment and he also examined a large number of their descendants. From the record cards for these animals I was able to obtain the body weights of a considerable number of rats belonging to the first six generations of the two inbred series. The body weight data for only the best of the animals reared during this period were used in the present case; records for all individuals that were noted as having deformed teeth or other malformations were excluded, as well as the records for those animals that were obviously runts.

Table 1 shows the body weight records for 92 males and for 73 females belonging in the first six generations of the A series of inbreds: table 2 gives similar data for 85 males and for 64 females of the B series. Only one record for each individual was obtained, i.e., the body weight at the time that the animal was killed.

The 'mean age in days,' as given in the first column of table 1 and of table 2, shows the median points in thirty periods that correspond to the ages at which the rats of the later inbred generations were weighed. For example, under the mean age '151 days' are grouped the body weights of all of the animals that were killed when they were from 136 to 165 days of age. As no records were taken of animals that were less than 105 days old the first age group given is that for 120 days.

TABLE 1

Showing the increase in the weight of the body with age for rats belonging in the first six generations of the A series of inbreds

MEAN AGE	MALES				FEMALES			
	Body weight in grams			Number of individuals	Body weight in grams			Number of individuals
	Average	Highest	Lowest		Average	Highest	Lowest	
days								
120	200.1	252	134	11	142.8	193	106	13
151	188.2	267	133	24	127.5	165	103	15
182	218.2	317	135	17	132.2	199	105	5
212	183.6	227	138	6	123.7	153	102	8
243	239.8	293	182	11	177.5	178	177	2
273	225.0	292	155	11	146.4	218	112	5
304	246.2	273	203	5	137.6	143	128	3
334	274.0	312	225	3	189.0	225	162	5
365		310		1*	164.8	215	143	5
395	273.0	294	269	3	177.1	191	151	6
425			.			166		1*
455		.			164 8	180	152	5
				92				73

* Record not used in constructing graph.

TABLE 2

Showing the increase in the weight of the body with age for rats belonging in the first six generations of the B series of inbreds

MEAN AGE	MALES				FEMALES			
	Body weight in grams			Number of individuals	Body weight in grams			Number of individuals
	Average	Highest	Lowest		Average	Highest	Lowest	
days								
120	205.8	266	142	24	149.7	170	136	4
151	210.6	257	141	16	149.4	211	113	11
182	241.6	345	146	15	136.0	185	109	5
212	235.7	315	157	12	119.2	173	100	4
243	270.5	304	238	4	169.7	189	149	4
273	215.6	241	185	6	161.0	193	131	4
304	233.7	287	174	4	161.7	189	138	12
334		249		1*	191.6	213	161	6
365	223.0	257	202	3	163.6	185	144	5
395					167.8	193	139	5
425								
455					172 2	213	145	4
				85				64

* Record not used in constructing graph.

A comparison of the corresponding data for the individuals of
the two series of inbreds is best made through the graphs in figure 1 constructed from the data for the average body weights at
different ages as given in table 1 and in table 2.

In figure 1 the graphs for the body growth of the males are
considerably higher than those for the females, since the male
rat, after reaching maturity, is normally a much heavier animal
than the female of like age.   Considering that only one record

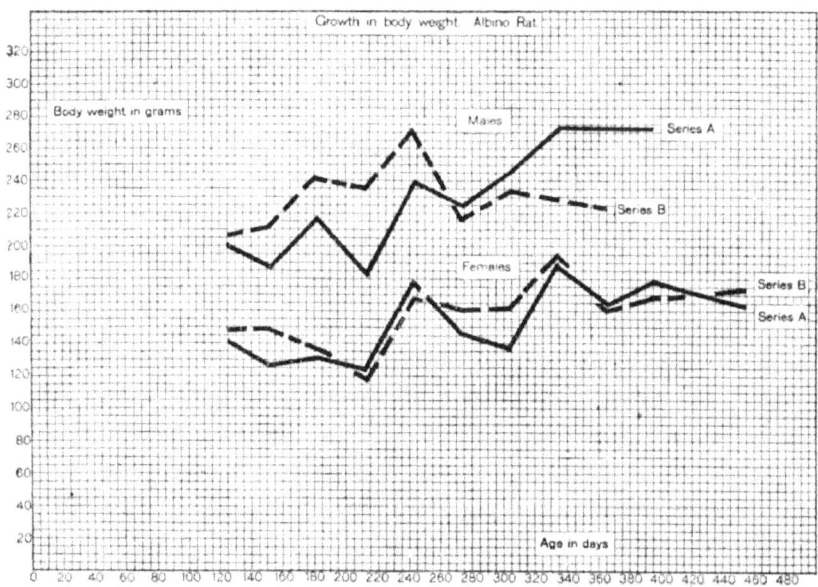

Fig. 1 Graphs showing the increase in the weight of the body with age for
males and females belonging in the first six generations of the two series of
inbred rats (data in table 1 and in table 2).

was taken for each animal the corresponding graphs for the two
series, especially those for the females, run remarkably close
together.   It is evident, therefore, that there was no significant
difference between the two series as regards the growth of the
individuals during the first six generations of inbreeding.

Table 3 gives the body weight data for all of the individuals
of the first six inbred generations for which records were taken
(a combination of the data in table 1 and in table 2).

TABLE 3

Showing the increase in the weight of the body with age for rats belonging in the first six generations of the inbred series. A combination of the data in table 1 and in table 2

MEAN AGE	MALES				FEMALES			
	Body weight in grams			Number of individuals	Body weight in grams			Number of individuals
	Average	Highest	Lowest		Average	Highest	Lowest	
days								
120	204.0	266	134	35	144.4	193	106	17
151	197.2	267	133	40	136.6	211	103	26
182	229.1	317	135	32	139.1	199	105	10
212	218.3	315	138	18	122.2	173	100	12
243	248.0	304	182	15	172.3	189	149	6
273	221.7	292	155	17	152.8	218	112	9
304	240.6	287	174	9	156.9	189	128	15
334	267.7	312	225	4	190.4	225	161	11
365	244.7	310	202	4	164.2	215	143	10
395	273.0	294	269	3	172.9	193	139	11
425						166		1*
455					168.1	213	145	9
				177				137

* Record not used in constructing graph.

A comparison of the data in table 3 with body weight data for a series of stock albino rats reared as controls for the inbred series after the change to 'scrap' diet had been made (table 13) will show to what extent malnutrition decreased the body size of the individuals in the first six inbred generations. Graphically this result is shown in figure 11 and in figure 12 (compare graph D with graph A and graph C).

Body weight data were taken, at the intervals stated, for 99 males and for 76 females belonging in the seventh to the fifteenth generation of the A series of inbreds. The average body weights of the males at different age periods are shown, by generations, in table 4: corresponding data for the females are given in table 5.

Data for 57 males and for 93 females belonging in the seventh . to the fifteenth generations of the B series of inbreds are shown according to generations in table 6 and in table 7.

TABLE 4

Showing, by generations, the average body weight at different ages of 99 males belonging in the seventh to the fifteenth generations of the A series of inbred rats

AGE	GENERATIONS								
	7	8	9	10	11	12	13	14	15
days	grams	grams	grams	grams	grams	grams	grams	grams	grams
13	18	18	18	18	16	19	19	18	18
30	43	45	46	41	44	44	49	47	42
60	150	77	134	129	102	127	134	123	123
90	201	151	199	214	177	184	186	193	189
120	242	196	249	264	227	229	244	233	204
151	274	252	287	289	257	249	273	259	235
182	306	286	308	315	279	273	291	273	256
212	312	315	330	299	288	293	311	283	266
243	329	341	342	318	302	295	317	293	286
273	355	352	364	305	321	312	327	304	300
304	357	345	334	310	314	319	333	302	303
334	368	368	376	305	327	316	339	307	307
·365	374	377	399	300	336	324	353	317	315
395	375	370	386	307	360	318	343	318	312
425	412	384	404	307	354	·334	355	330	321
455	424	383	403	324	343	339	344	322	320
Number of rats weighed..........	6	6	9	14	12	10	15	15	12

TABLE 5

Showing, by generations, the average body weight at different ages of 76 females belonging in the seventh to the fifteenth generations of the A series of inbred rats

AGE	GENERATIONS								
	7	8	9	10	11	12	13	14	15
days	grams	grams	grams	grams	grams	grams	grams	grams	grams
13	16	15	16	15	15	18	17	17	18
30	40	37	42	37	41	41	45	45	40
60	126	73	116	107	90	102	110	108	105
90	160	117	156	134	139	145	157	162	144
120	172	131	181	184	181	168	172	171	163
151	190	163	205	189	182	187	192	185	183
182	190	182	217	208	189	195	196	204	203
212	187	193	211	193	184	218	215	209	211
243	189	200	223	202	200	203	216	214	218
273	187	219	229	223	192	226	218	217	222
304		221	221	236	209	239	221	222	222
334		217	220	242	222	244	218	223	227
365		252	256	256	223	238	220	226	226
395		232		253	226	253	227	228	224
425		252		244	260*	268*	230	226	219
455		241		257	257*	243	224	219	217
Number of rats weighed..........	4	9	7	8	10	8	9	10	11

* One record only.

TABLE 6

Showing, by generations, the average body weight at different ages of 57 males belonging in the seventh to the fifteenth generations of the B series of inbred rats

AGE	GENERATIONS								
	7	8	9	10	11	12	13	14	15
days	grams	grams	grams	grams	grams	grams	grams	grams	grams
13	21	21	20	18	19	19	19	20	20
30	49	49	53	49	40	47	50	52	49
60	170	158	140	172	149	140	144	132	145
90	244	197	211	217	220	212	193	210	201
120	277	280	253	265	262	240	225	255	233
151	322	293	288	291	300	264	274	279	267
182	344	328	340	334	330	278	283	300	288
212	376	331	360	340	353	290	297	305	303
243	372	348	367	336	368	286	303	302	308
273	454*	356	371*	316	367	306	309	311	311
304	477*	361	380*	335	408	313	313	322	311
334		361	372*		381	337	303*	330	321
365		364	345*	333	373	328	326*	362	336
395		369		365*	365	336*	334*	351	339*
425		365*			353			357*	339*
455					347*			343*	330*
Number of rats weighed.........	3	7	3	5	9	8	6	7	9

* One record only.

TABLE 7

Showing, by generations, the average body weight at different ages of 93 females belonging in the seventh to the fifteenth generations of the B series of inbred rats

AGE	GENERATIONS								
	7	8	9	10	11	12	13	14	15
days	grams	grams	grams	grams	grams	grams	grams	grams	grams
13	18	18	17	17	18	18	18	18	18
30	45	40	50	48	39	39	46	48	45
60	135	119	122	125	113	120	104	104	107
90	187	168	168	170	165	149	147	161	151
120	187	181	180	174	182	167	169	183	172
151	204	200	200	199	191	186	195	196	191
182	213	196	205	208	200	195	196	214	202
212	216	213	199	217	202	204	204	210	206
243	208	216	211	217	225	212	214	216	201
273	243	196*	218	212*	236	208	220	219	214
304	246	201*	217	239	243	207	225	223	217
334	235	204*	239		255	211	243	237	224
365	249	210*	242	262*	249	215	238	241	229
395	301*		241	275*	241	212	226	239	229
425	323*			261*	238	207		239	234
455	317*			280*	244			231	235
Number of rats weighed.........	7	11	5	11	12	10	9	12	16

* One record only.

Tables 4 to 7 are inserted chiefly for reference, although they bring out two important facts more clearly, perhaps, than do any of the other tables. The data, as given in these tables, show that rats belonging to the earlier generations of the inbred series did not live as long, as a rule, as did the individuals in the later generations. This was particularly noticeable in the individuals of the B series. Up to the twelfth generation only three rats in the B series (two females and one male) lived to the age of 455 days; in subsequent generations many individuals lived for the entire weighing period of fifteen months, and some of them were kept until they were nearly two years old. One who believes with Crampe and Ritzema-Bos that continued inbreeding necessarily lessens vitality and so shortens the life of the individual meets here with the seemingly paradoxical fact that the animals that belonged to the later inbred generations outlived those that belonged to the earlier generations. In this experiment the use of only the most vigorous animals for breeding purposes has seemingly overcome any tendency that inbreeding might have to shorten the life of the individuals.

The second point of interest brought out by tables 4 to 7 is that in each generation of the two inbred series the average body weight of the males exceeded that of the females at every age for which records were taken. At birth the male albino rat is slightly heavier than the female, whether the animals belong to a stock or to an inbred strain (King, '15 b). Data for the growth in body weight of the albino rat, as recorded by Donaldson ('06), show that as early as the seventh day after birth the growth of the female is more vigorous than that of the male, and that the female is, as a rule, a relatively heavier animal than the male up to about fifty-five days of age. Ferry's ('13) growth data for the albino rat (Donaldson, '15; table 65) confirm Donaldson's findings. In Jackson's ('13) data for the albino rat, "the excess of average weight was invariably in favor of the male at birth, and also in the majority of cases at all succeeding ages;" while the records obtained by Hoskins ('16) show that the albino female is a heavier animal than the male only at the age of about six weeks. In a series of stock albino rats reared as controls for the inbred series (King, '15 a) the majority of the males exceeded the females in body weight at each weighing period. As no records were taken when the animals were just six weeks old, the relative size of the two sexes at this age was not determined.

At thirteen days of age, as tables 4 to 7 show, the average body weight of the inbred males was only one gram more than that of the females. Such a slight difference as this would be negligible, considering the possible error in the weighings due to the varying amount of food in the alimentary tract, save for the fact that it is found in every generation group. In some litters the females, were, on the average, heavier than the males at thirteen and also at thirty days of age, and in a few instances the weight of the largest female surpassed that of the smallest male even when the animals were sixty days old. Although the albino female, whether stock or inbred, has a relatively smaller birth weight than the male, she soon comes to have a body weight that is very nearly equal to that of the males in all cases, and often exceeds it. Even though the absolute body weights are less at any given age, therefore, the female grows more vigorously than the male during the first few weeks of postnatal life. An early acceleration in the growth of the female also occurs in the guinea-pig (Minot, '91; Castle, '16), and it finds a parallel in man, as Donaldson ('12) has pointed out, since during a certain phase of development, when children are from thirteen to sixteen years old, the average weight of girls is greater than that of boys; while at all other ages boys, as a rule, are the heavier.

The relative growth in body weight of males and of females in corresponding generations of the two inbred series, or in succeeding generations of the same series, can be determined by referring to the data in tables 4 to 7. For a comparative study of the body growth of the individuals in the two series it seemed advisable to combine the data for three succeeding generations. Data for the A series of inbreds are given in table 8.

The growth graphs shown in figure 2 were constructed from the data for the males of the A series, as given in table 1 and

TABLE 8

Showing the average body weights at different ages of inbred rats of the A series separated into three groups according to the generation to which the individuals belonged

AGE	MALES			FEMALES		
	Generations 7–9	Generations 10–12	Generations 13–15	Generations 7–9	Generations 10–12	Generations 13–15
days	*grams*	*grams*	*grams*	*grams*	*grams*	*grams*
13	18.1	17.5	18.4	15.5	16.0	17.5
30	49.4	43.4	43.6	42.5	40.0	43.2
60	122.3	118.3	127.0	98.9	98.5	107.6
90	186.0	193.2	188.9	137.5	157.9	153.6
120	230.6	242.5	228.1	160.5	178.5	168.3
151	269.5	267.9	256.9	181.5	191.0	186.2
182	304.3	290.4	276.7	192.2	195.3	201.9
212	320.2	293.2	287.1	189.8	196.2	211.2
243	337.1	305.9	298.3	205.5	201.7	216.0
273	357.4	312.5	307.7	214.7	215.1	219.4
304	349.5	314.5	313.7	221.1	219.4	221.7
334	369.0	315.4	318.4	227.1	234.9	223.5
365	379.2	318.1	327.1	252.2	235.2	224.7
395	375.6	323.6	323.6	232.0	239.5	226.4
425	399.4	327.0	332.0	252.5	254.0	223.8
455	403.5	336.0	333.2	241.0*	251.6	219.0

* One record only.

in table 8. In this, as in some of the other figures, the space between the graphs was slightly widened, where properly the lines would run very close together or overlap, in order that the course of each graph might be clearly followed.

Figure 2 shows that the body growth of the males in the three generation groups of the A series of inbreds progressed at about the same average rate until the animals were 150 days of age, as is indicated by the position of graphs A to C. At this point there was a marked acceleration in the growth of the males belonging to the group comprising the seventh to the ninth generation (graph A) which continued until the end of the weighing period. Graph D, representing the growth of the males of the A series during the first six generations, begins at the 120-day period, as no younger rats belonging to these generations were weighed. It seems almost incredible that this graph can represent the adult body size of the progenitors of the animals whose growth is indicated by graph A, since rarely do we find a group of mammals having in the adult state a body size so much greater than that of its immediate ancestors.

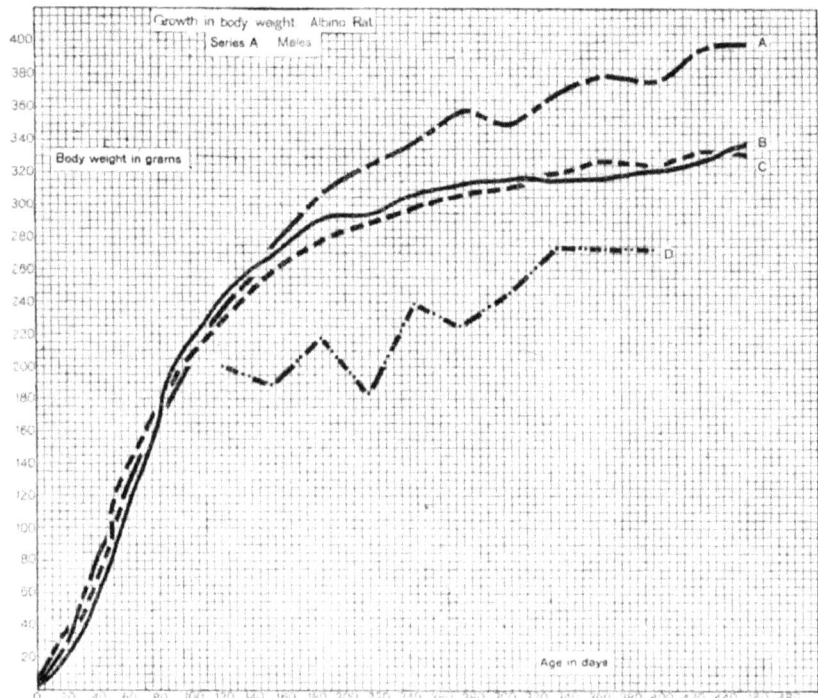

Fig. 2　Graphs showing the increase in the weight of the body with age for males belonging to various generation groups of the A series of inbreds (data in table 1 and in table 8).　A, graph for males of the seventh to the ninth generations inclusive; B, graph for the males of the tenth to the twelfth generation inclusive; C, graph for the males of the thirteenth to the fifteenth generations inclusive; D, graph for the males of the first six inbred generations.

Graphs showing the growth in body weight of females belonging to the three generation groups of the A series are shown in figure 3 (data in table 1 and in table 8).

The females of the first six generations of the A series were very much smaller than the females of the later generations at every age for which records were taken, as the position of graph D in figure 3 clearly shows.　There was practically no difference in the rate or in the extent of the body growth in the groups comprising the seventh to the fifteenth generations, as graphs A, B and C cross and recross each other at various points and run as close together as would any set of graphs constructed from the data for different series of individuals.

Data for the growth in body weight of the individuals of the B series, arranged in groups of three generations each, are given in table 9.

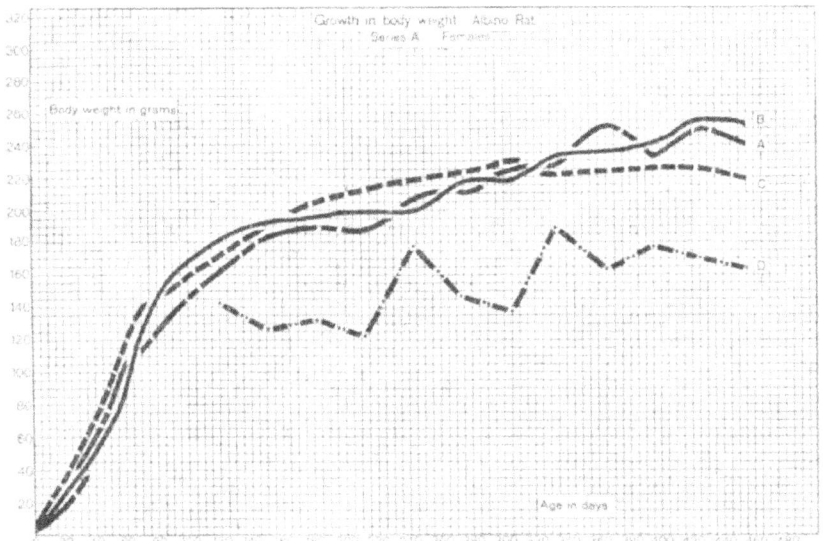

Fig. 3 Graphs showing the increase in the weight of the body with age for females belonging to various generation groups of the A series of inbreds (data in table 1 and in table 8: lettering as in figure 2).

From the average body weights at different ages of the males of the B series of inbreds, as given in table 2 and in table 9, the graphs in figure 4 have been constructed.

In the B series, from the first weighing until the last, there was a marked difference in the body weights of the males in the four generation groups, since all of the graphs in figure 4 are distinctly separated except at one point (365-day period). Graph A runs higher than any of the other graphs from the beginning until the end of its course, thus indicating that in this series of inbreds also the males of the seventh to the ninth generation were heavier animals than the males in the later generation groups. Males in the first six generations of the B series

TABLE 9

Showing the average body weights at different ages of inbred rats of the B series separated into three groups according to the generation to which the individuals belonged

AGE	MALES			FEMALES		
	Generations 7–9	Generations 10–12	Generations 13–15	Generations 7–9	Generations 10–12	Generations 13–15
days	*grams*	*grams*	*grams*	*grams*	*grams*	*grams*
13	21.0	19.1	19.8	17.9	17.7	18.2
30	50.0	47.2	50.0	43.8	44.8	46.2
60	156.9	151.4	140.7	120.3	119.6	105.4
90	218.9	213.4	201.6	172.8	163.3	153.6
120	273.6	259.7	237.6	182.8	175.9	174.6
151	299.0	284.3	271.6	201.3	190.4	193.5
182	335.3	309.6	289.7	205.3	200.6	205.4
212	347.5	323.4	301.7	210.4	209.5	206.7
243	358.3	321.4	304.9	211.6	217.7	210.5
273	372.2	330.4	310.2	222.2	217.1	217.1
304	380.5	353.2	316.5	225.6	222.4	221.3
334	363.1	359.0	323.1	230.6	228.6	230.8
365	360.4	342.4	344.2	238.4	229.3	234.3
395	369.0	357.7	342.2	261.3	236.4	232.6
425	365.0*	353.0	348.0	323.0*	230.4	236.4
455		347.0*	331.5	317.0*	256.3	233.2

* One record only.

TABLE 10

Showing the average body weights at different ages of inbred rats of the combined series (A, B) separated into three groups according to the generation to which the individuals belonged

AGE	MALES			FEMALES		
	Generations 7–9	Generations 10–12	Generations 13–15	Generations 7–9	Generations 10–12	Generations 13–15
days	*grams*	*grams*	*grams*	*grams*	*grams*	*grams*
13	19.2	18.1	18.9	16.8	16.9	17.8
30	49.6	44.9	47.6	43.0	42.8	45.1
60	135.5	131.8	131.7	112.7	110.7	106.4
90	195.6	200.9	193.1	154.1	160.7	153.6
120	248.6	250.5	231.4	172.1	177.2	171.8
151	281.1	274.3	262.2	191.2	190.7	190.2
182	315.6	298.0	279.5	199.2	198.5	203.5
212	328.7	305.7	291.6	203.4	202.2	208.3
243	343.9	311.4	300.0	208.6	209.4	213.3
273	361.6	316.1	308.5	217.4	215.8	218.2
304	359.8	322.2	314.2	222.8	220.8	221.5
334	367.3	324.1	319.3	228.7	232.8	226.6
365	374.0	325.5	329.2	243.7	232.4	228.6
395	374.4	331.4	325.3	246.6	238.3	229.0
425	396.2	330.5	333.4	276.0	240.8	229.6
455	403.5	336.7	333.0	279.0	253.3	224.6

were quite as inferior to their descendants in body size as were the males in the corresponding generations of the A series (graph D).

The growth in body weight of various groups of females belonging in the B series is shown graphically in figure 5 (data in table 2 and in table 9).

Graphs A to C in figure 5 were drawn so that the lines are distinct. These graphs should lie very close together and

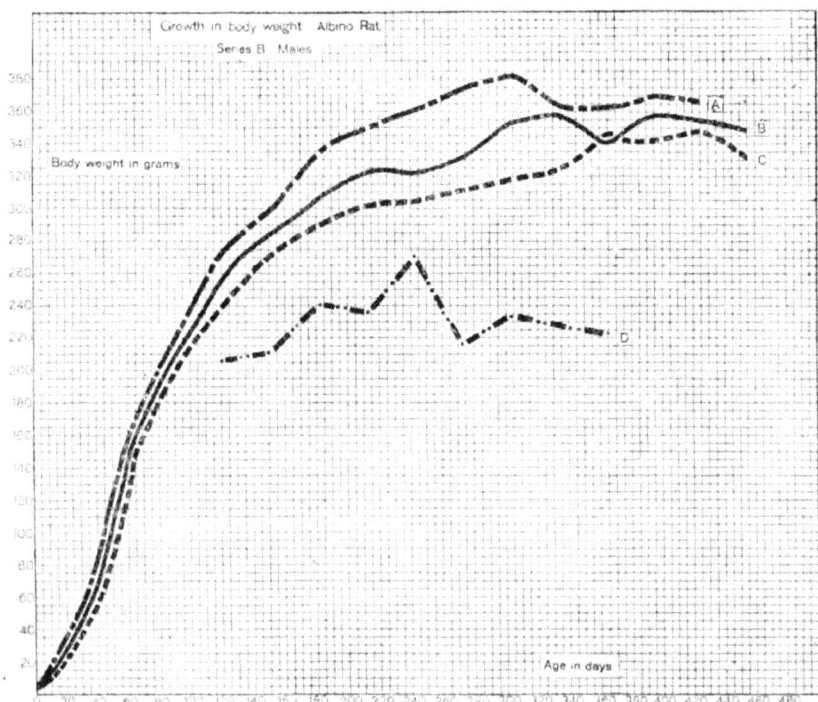

Fig. 4 Graphs showing the increase in the weight of the body with age for males belonging to various generation groups of the B series of inbreds (data in table 2 and in table 9: lettering as in figure 2).

overlap in many places, since the data in table 9 shows that the actual differences between the corresponding body weights of the various groups are very small. The position of graph D indicates that adult females of the first six generations were very much smaller animals than their descendants of like age, as was the case in the A series also.

The data given in table 8 and in table 9 have been combined in table 10.

Figure 6 shows graphs for the weight increase with age in all of the inbred males for which weight records were taken (data in table 3 and in table 10).

As the position of the graphs in figure 6 show, males of the seventh to the ninth inbred generations (graph A) were animals of unusually large size and they were considerably heavier, after reaching maturity, than their own progeny of like age. At the 365 day period the space between graph A and graph B indicates a difference of about 50 grams in the average body

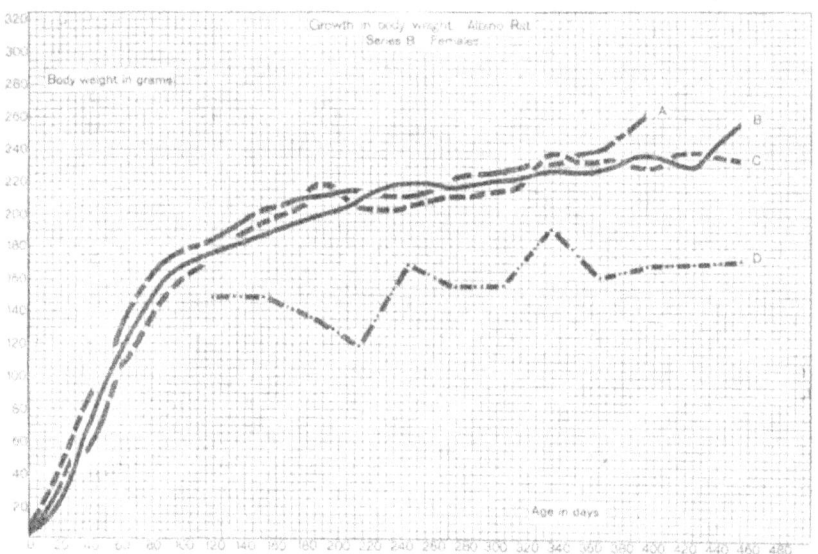

Fig. 5 Graphs showing the increase in the weight of the body with age for females belonging to various generation groups of the B series of inbreds (data in table 2 and in table 9: lettering as in figure 2).

weights of the two groups of animals. From the tenth generation on the course of growth in body weight was practically the same in all inbred males, as is shown by the fact that graph B and graph C run very close together throughout their entire length.

Figure 7 shows graphs for the body weight increase in the series of inbred females (data in table 3 and in table 10).

The relative position of the graphs in figure 7 show that, with the exception of the animals in the first six inbred genera-

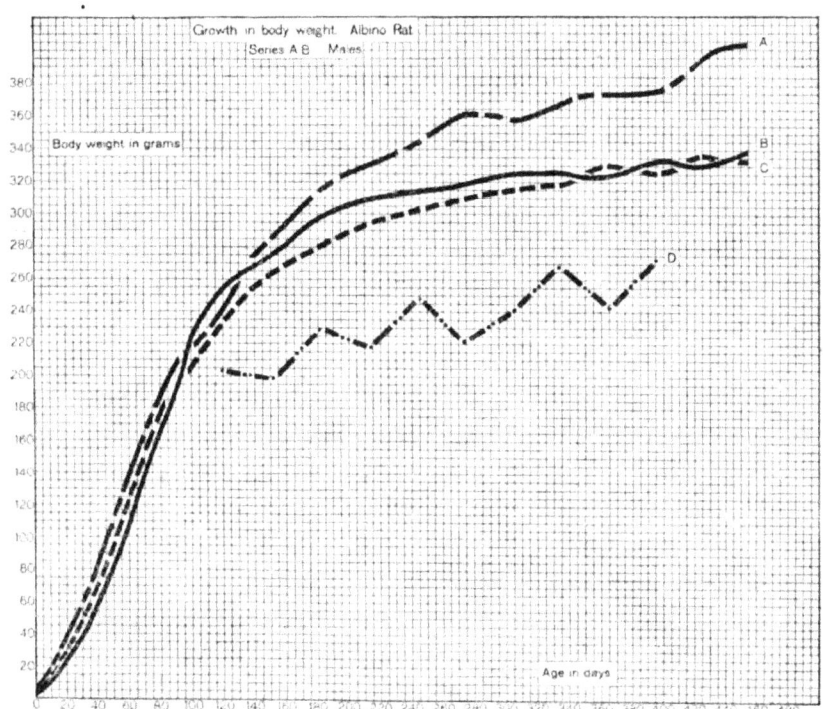

Fig. 6 Graphs showing the increase in the weight of the body with age for males belonging to various generation groups of the two series combined (A, B). Data in table 3 and in table 10: lettering as in figure 2.

Fig. 7 Graphs showing the increase in the weight of the body with age for females belonging to various generation groups of the two series combined (A, B). Data in table 3 and in table 10: lettering as in figure 2.

23

tions (graph D), the body increase with age was about the same
in all the various generation groups of inbred females up to the
time that the animals were one year old. After this age the
females in the group comprising the seventh to the ninth generations were somewhat heavier than the other females, but their
excess in body weight was very much less than that in the corresponding groups of males (fig. 6).

On referring to the data in table 4 to table 7 it is found that
in both series of inbreds the average body weights of the animals
in the seventh, eighth and ninth generations were greater than
those of animals in the tenth and subsequent generations. In
the seventh generation, particularly, females as well as males
were exceptionally heavy animals at all ages for which records
were taken. The largest males yet obtained belonged in the
seventh generation of the A series of inbreds; these rats weighed
at their maximum 460, 482, and 511 grams respectively. The
largest female in the series was a member of the seventh generation of the B series of inbreds, and she weighed 323 grams when
she was 425 days old. The probable cause for the unusual
weight of the animals in the seventh to the ninth inbred generations will be considered later.

Data showing the range in variation and the averages for the
body weights at different ages of the weighed individuals in the
A series of inbreds (seventh to fifteenth generations only) are
given in table 11. Similar data for the rats of the B series are
shown in table 12.

The graphs in figure 8 were drawn to show the relative growth
in body weight of the males belonging to the two inbred series
(data in table 11 and in table 12).

The males of the B series of inbreds had a heavier average
body weight than the males of the A series up to the last weighing period (455 days), as the position of the graphs in figure 8
show. The crossing of the graphs at the end has no significance,
since the largest males of the B series died before the final
weighing.

Graphs showing the growth in body weight of the females in
the two inbred series are shown in figure 9: the data for these
graphs are given in table 11 and in table 12.

TABLE 11

Showing the increase in the weight of the body with age for 99 male and for 76 female rats belonging in the seventh to the fifteenth generations of the A series of inbreds

AGE	MALES				FEMALES			
	Body weight in grams			Number of individuals	Body weight in grams			Number of individuals
	Average	Highest	Lowest		Average	Highest	Lowest	
days								
13	17.9	21	15	99	16.1	22	14	76
30	45.6	82	35	99	41.8	67	33	76
60	123.0	176	72	95	102.3	159	71	74
90	189.8	248	120	99	150.3	193	93	63
120	233.9	305	174	94	170.0	212	133	63
151	263.4	360	204	94	187.6	224	157	65
182	287.4	367	216	91	197.5	236	167	62
212	296.8	420	238	85	202.8	248	158	59
243	309.5	419	234	84	209.9	245	171	56
273	322.6	435	265	74	217.0	251	167	53
304	321.1	415	267	70	221.0	257	181	42
334	328.9	415	268	63	227.2	268	196	37
365	336.2	428	289	59	229.8	282	198	35
395	335.9	433	283	51	230.3	274	206	31
425	345.9	448	281	44	232.2	268	206	21
455	361.3	473	306	25	228.3	257	182	20

TABLE 12

Showing the increase in the weight of the body with age for 57 male and for 93 female rats belonging in the seventh to the fifteenth generations of the B series of inbreds

AGE	MALES				FEMALES			
	Body weight in grams			Number of individuals	Body weight in grams			Number of individuals
	Average	Highest	Lowest		Average	Highest	Lowest	
days								
13	19.8	27	16	57	18.0	25	14	93
30	48.6	75	35	57	45.1	72	30	93
60	148.5	190	110	57	115.2	147	80	93
90	210.1	266	149	57	161.1	197	117	68
120	254.1	331	218	55	176.8	198	134	73
151	282.9	365	232	55	193.4	229	156	67
182	307.5	407	253	52	213.7	241	168	70
212	319.4	410	254	44	208.2	247	174	60
243	323.1	408	259	39	212.5	261	179	54
273	330.7	454	275	32	213.3	278	177	47
304	344.3	477	290	28	221.1	281	179	40
334	344.4	385	299	18	230.3	290	179	26
365	348.2	376	298	17	233.7	276	194	29
395	355.1	374	333	11	237.2	301	206	23
425	353.2	365	339	5	239.4	323	210	19
455	336.6	347	320	3	245.0	317	212	13

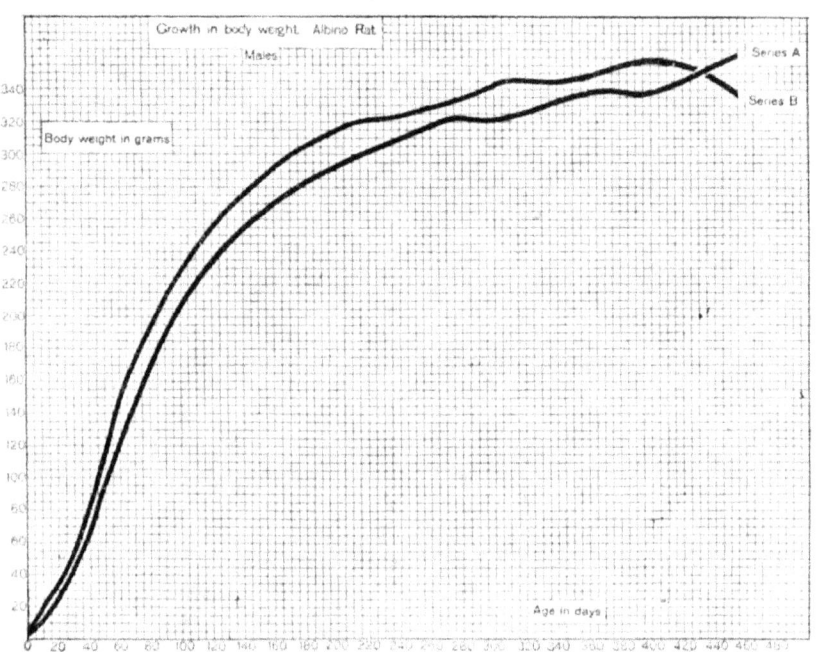

Fig. 8 Graphs showing the increase in the weight of the body with age for males belonging in the seventh to the fifteeth generations of the two series of inbreds (data in table 11 and in table 12).

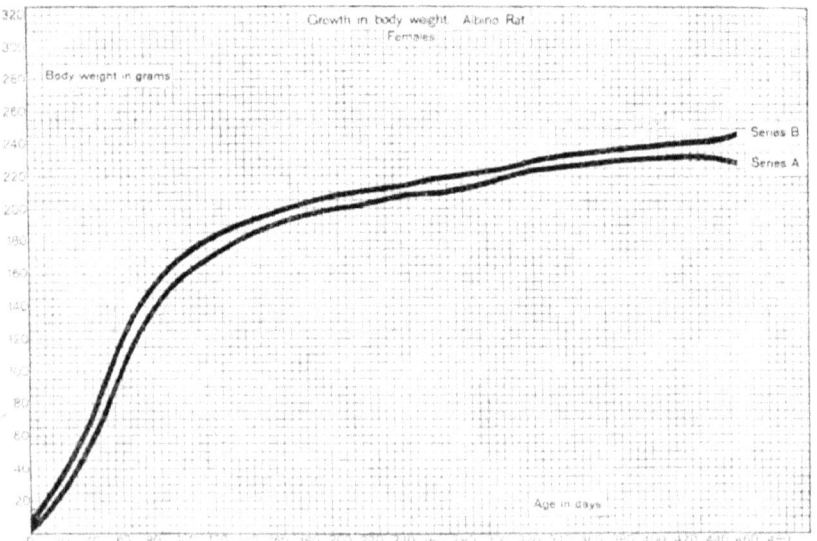

Fig. 9 Graphs showing the increase in the weight of the body with age for females belonging in the seventh to the fifteenth generations of the two series of inbreds (data in table 11 and in table 12).

26

The graphs in figure 9 run close together throughout their entire course, but the graph for the B series is higher at all points than that for the A series. It appears, therefore, that growth in body weight was somewhat more vigorous in the females of the B series than in those of the A series.

It seems a rather significant fact that in both figure 8 and figure 9 the graphs run parallel from the beginning until the end of their course; they do not cross and recross at various points, as one might expect would be the case with graphs for two series of rats from the same ancestral stock that were reared simultaneously under the same environmental conditions. Female B and her mate, the ancestors of the B series of inbreds, were heavier animals at the time that they were killed than were the progenitors of the A series of inbreds, female A and her mate. Body weight in the rat is so dependent on physical condition, however, that a single weighing of the animals when they were at an advanced age would not necessarily give a true idea of the relative size of the animals at an earlier age period. The difference in the size of the two pairs of rats with which the experiments were started, together with the fact that after the sixth generation the descendants of female B were relatively heavier animals than the descendants of female A, point to the conclusion that the difference in the size of the animals in the two series was not due to chance or to environment, but that it was dependent in some way upon the inheritance of genetic factors for growth.

Table 13 gives the body weight data for the total of 156 males and 169 females in the seventh to the fifteenth inbred generations for which weight records were taken (a combination of the data in table 11 and in table 12).

Table 13 brings out one fact of interest: the average body weight of the male inbred rats increased with age up to the end of the weighing period when it was 358.7 grams; the average body weight of the females was at its maximum at the 425 day period, and then fell off slightly at the final weighing. Individual rats show a pronounced difference as regards the time that they attain their maximum body weight and, as a rule, the

TABLE 13

Showing the increase in the weight of the body with age for 156 male and 169 female rats belonging in the seventh to the fifteenth generations of the two series (A, B). A combination of the data in table 11 and in table 12

AGE	MALES				FEMALES			
	Body weight in grams			Number of individ-uals	Body weight in grams			Number of individ-uals
	Average	Highest	Lowest		Average	Highest	Lowest	
days								
13	18.6	27	15	156	17.0	25	14	169
30	46.5	82	35	156	43.6	72	30	169
60	132.6	190	72	152	109.4	159	71	167
90	197.3	266	120	156	155.9	197	93	131
120	241.4	331	174	149	173.7	212	133	136
151	270.6	365	204	149	190.5	229	156	132
182	294.7	407	216	143	200.8	241	167	132
212	304.5	420	238	129	205.8	248	158	119
243	313.8	419	234	123	211.2	261	171	110
273	325.0	454	254	106	215.3	278	167	100
304	327.7	477	267	98	221.0	281	179	82
334	332.3	415	268	81	228.4	290	179	63
365	338.8	428	289	76	231.5	282	194	64
395	339.3	433	283	62	233.2	301	206	54
425	346.6	448	281	49	235.6	323	206	40
455	358.7	473	306	28	234.9	317	182	33

females reach this point at an earlier age than do the males. The records for these inbred rats show that in many cases the maximum body weight in both sexes came when the animals were seven or eight months of age, at which time they were at the height of their reproductive activity (King, '16), and then gradually decreased; other rats increased steadily in body weight until they were sixteen or even eighteen months old. Autopsies made on a considerable number of unusually large rats indicate that the later weight increase is chiefly an accumulation of adipose tissue and is not, therefore, to be considered as true growth.

Under the very favorable climatic conditions of California, Slonaker ('12 a) found that the albino rat reaches its maximum body weight, as a rule, by the age of fifteen months. In one series of animals whose weight records were taken at frequent intervals from the time of weaning until natural death, Slonaker found that the ten males attained their average maximum body weight of 247.5 grams when they were 434 days of age, and that the six females reached their maximum weight of 151.8 grams somewhat earlier than did the males i.e. at 375 days of age. After the maximum was reached there was a slow but steady

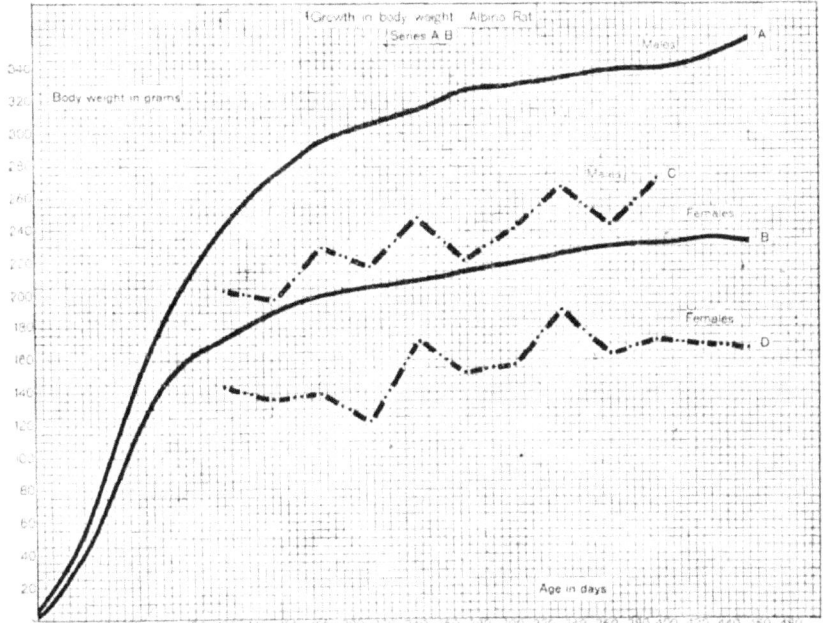

Fig. 10 Graphs showing the increase in the weight of the body with age for males and for females belonging in various generations of the two series combined (A, B). Data in table 3 and in table 13. A and B, graphs for individuals in the seventh to the fifteenth generations inclusive; C and D, graphs for animals in the first six inbred generations.

decline in body weight, although the animals appeared to be in good physical condition, and some of them lived to be nearly four years old.

The graphs in figure 10 show the weight increase with age for the total number of males and females in the two inbred series for which weight records were obtained (data in table 3 and in table 13).

The graphs in figure 10 show in a very striking manner the
great difference between the size of the rats in the first six inbred generations and those in the later generations. At the
300 day period the space between graph A and graph C represents a difference of 87 grams in the average body weights of the
two groups of males. Females of the earlier generations (graph
D) were likewise far inferior in body size to the females of subsequent generations (graph B), and at 300 days of age the space
between graph B and graph D indicates a difference of 64 grams
in favor of the females in the later generation group.

The earliest data on the growth in body weight of the albino
rat are those of Donaldson ('06) who studied the growth changes
in a series of animals reared at The University of Chicago. Other
investigators, Jackson, Slonaker, Ferry and Hoskins have published records for the growth in body weight of various series of
albino rats reared under different environmental conditions.
All of the latter data agree, in the main, with those of Donaldson, although as the rat is very responsive to external conditions
some series of records show more rapid and vigorous growth
than others.

Donaldson's growth graphs for albino rats may be taken as
representing the average run of stock animals. His graph for
the males is reproduced as graph A in figure 11, and that for
the females is shown as graph A in figure 12.

As controls for the present series of inbred albino rats thirteen litters of stock albinos, comprising fifty males and fifty
females, were reared in The Wistar Institute animal colony
under the same environmental and nutritive conditions as the
later generations of the inbred series. In selecting litters for
controls care was taken to pick out only those in which the young
were large and vigorous at birth. The animals chosen, therefore, represent the best, not the average, stock in The Wistar
colony. The average body weights at various ages of the males
and females in this selected group of stock albinos, reproduced
from table 3 of a previous publication (King, '15 a), are given
in table 14.

Figures 11 and 12 show graphs for weight increase with age in two series of stock albinos (graphs A and C), together with graphs representing the growth of animals in the inbred series (graphs B and D). The graphs for the various groups of males are given in figure 11.

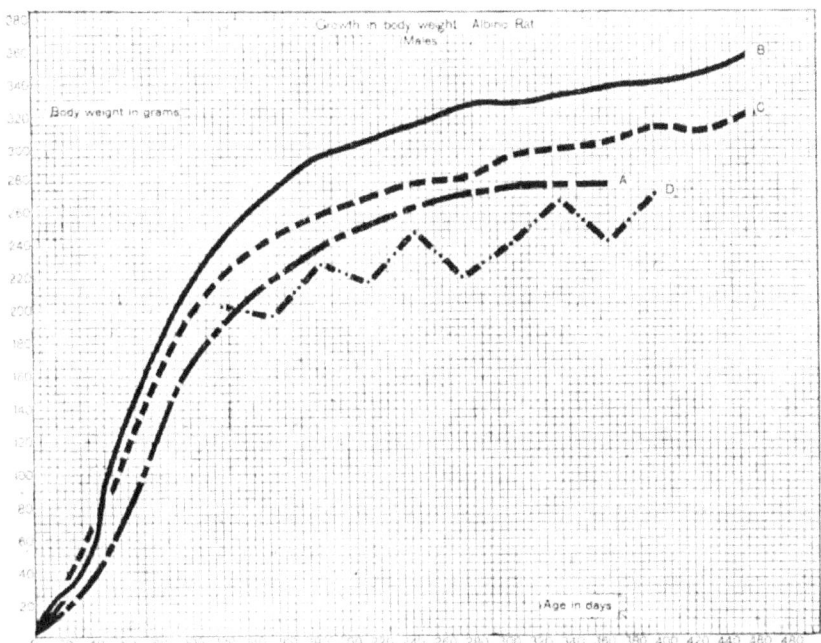

Fig. 11 Graphs showing the increase in the weight of the body with age for different series of male albino rats. A, graph constructed from Donaldson's data for stock albinos; B, graph for males belonging in the seventh to the fifteenth generations of the two series of inbreds combined; C, graph constructed from data for a selected series of stock albinos used as controls for the inbred strain; D, graph for males belonging in the first six generations of the two series combined.

In figure 11 graph B, representing the growth of the males in the inbred series, runs higher than Donaldson's graph for stock albinos (A) from the beginning until the end of its course. At the 243 day period the space between these graphs represents a difference of about 18 per cent in the average body weights of the two series of animals. At all points, except the thirty day period, graph B is higher than graph C which shows the body growth of the selected stock males reared as controls for the inbred group. Data in table 13 and in table 14 show that at 13 days of age the inbred males weighed, on the average, 1.4 grams more than did the stock males of the same age, but that at thirty days of age the average body weight of the stock group was two grams more than that of the inbreds; after this age inbred males increased in body weight much more rapidly than did the stock animals. When the rats were at their prime, at eight months of age, inbred males were about 12 per cent heavier than the males of the control series.

The above analysis of data shows that not only were inbred males of the seventh to the fifteenth generation much heavier than the general run of stock animals at any given age, but that they were also larger, except at the thirty day period, than the selected stock controls reared under similar environmental conditions.

Graphs showing the weight increase with age for various groups of stock and inbred females are given in figure 12.

In figure 12 graph A, which indicates the body growth of the females of Donaldson's series of stock albinos, is not strictly comparable to the other graphs, since it was constructed from the actual weight data for unmated females only up to the ninety day period, beyond this point the data used were those of unmated females corrected to accord with the weights of breeding females as calculated by Watson's ('05) formula: all other graphs were based on the actual weight data of breeding females. Graph A runs lower than either the graph for the inbred females (B) or that for the stock controls (C) during the period in which the actual weight records were used in constructing the graph, but later it is considerably higher than the other graphs. It would seem as if the corrective factor introduced in Watson's formula was much too high, since no series of actual weight records for the albino rat yet reported comes up to the standard required by Donaldson's graph.

In figure 12 the graph for the growth of inbred females (B) and that for the females in the control series (C) run very close together throughout their entire length. From the thirty to

the sixty day period graph C is a little above graph B, but at
all other points graph B is the higher of the two.   At the 243
day period the space between the graphs represents a difference
of only 0.76 per cent in the average body weights of the two
groups of females, although the inbred females were 3.7 per
cent heavier than the stock females when the animals reached
one year of age.

These results indicate that the rate and the extent of the
growth in inbred females was about the same as that of the females in the selected stock controls during the adolescent period,

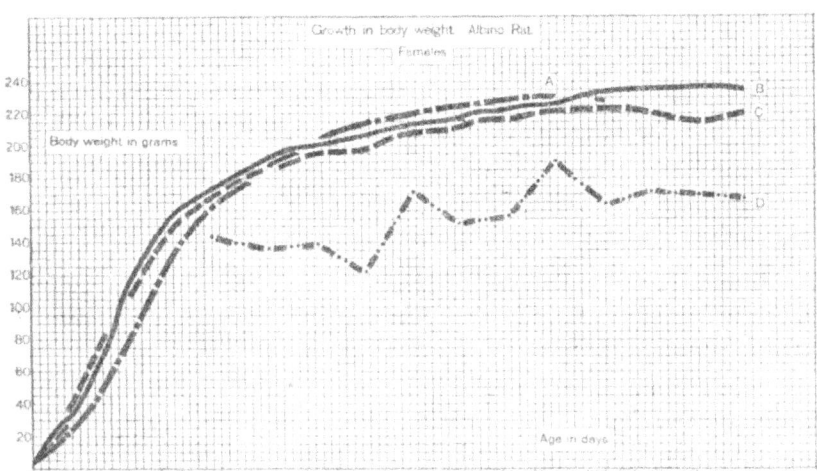

Fig. 12 Graphs showing the increase in the weight of the body with age for
different series of female rats (lettering as in figure 11).

but that in the adult state the inbred females tended to be
slightly heavier than stock females of the same age.

As shown in various tables (4–10) and in several figures (nos.
2, 4, 6) rats belonging to the later inbred generations were not
as heavy at any given age as the animals in the seventh to the
ninth inbred generations.   One naturally asks whether inbreeding has lessened the growth impulse and impaired the vitality
after many generations so that these animals are inferior in body
size to normal stock animals of like age.   In order to answer
this question, data showing the increase in the body weight

TABLE 14

Showing the average increase in the body weight with age for two series of albino rats: (1) rats belonging to the fifteenth generation of the two inbred series (A, B); (2) selected stock rats. Both series of rats were reared under similar environmental conditions

AGE	AVERAGE BODY WEIGHT IN GRAMS		AVERAGE BODY WEIGHT IN GRAMS	
	Males		Females	
	Inbreds 21 rats	Stock 50 rats	Inbreds 27 rats	Stock 50 rats
days				
13	19.1	17.2	18.1	15.7
30	44.9	48.5	42.8	45.7
60	132.3	122.9	106.2	107.1
90	184.9	184.8	147.9	148.0
120	216.1	223.2	168.2	173.4
151	249.1	244.8	184.1	186.3
182	269.0	258.4	202.4	196.5
212	278.3	268.0	208.0	197.3
243	291.3	279.7	210.6	209.6
273	302.0	280.9	218.0	210.8
304	305.3	296.1	219.8	219.1
334	309.8	300.8	225.4	222.4
365	318.8	306.1	227.3	223.1
395	315.8	314.1	226.2	220.5
425	323.2	312.2	227.4	215.8
455	319.7	323.9	226.3	220.2

with age in rats belonging to the fifteenth generation of the inbred series (A, B), with corresponding data for the animals in the Wistar stock control series are given in table 14.

Data for the two series of male rats, as given in table 14, are presented in the form of graphs in figure 13.

In figure 13 the graphs run close together and cross and recross until the 150 day period. At this point the graph for the inbred males mounts above that for the stock animals and subsequently maintains this position until the end, where the stock graph is the higher, since at this age the stock males were 1.3 per cent heavier, on the average, than the inbred males. At the eight months period the space between the graphs indicates a difference of 4.1 per cent in favor of the inbreds.

The majority of the inbred rats in the fifteenth generation were handicapped in their growth by being born in the summer:

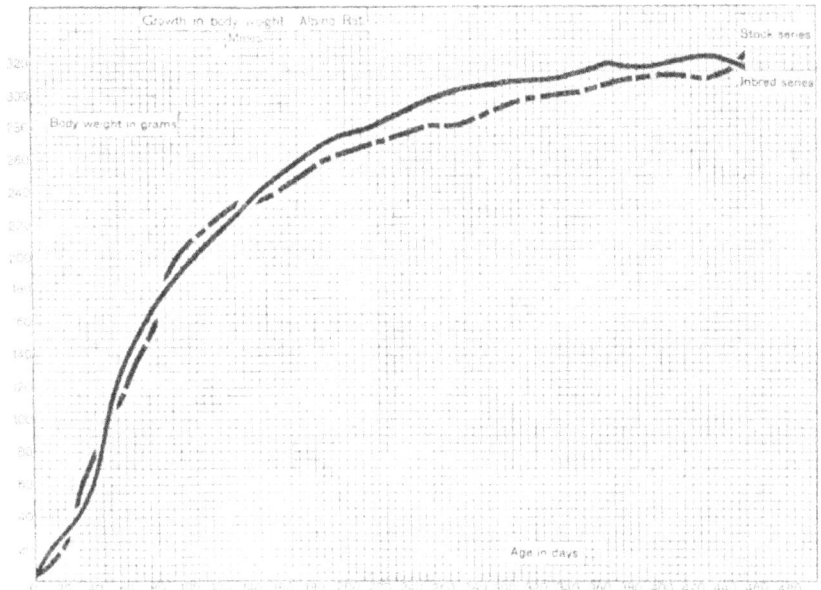

Fig. 13 Graphs showing the increase in the weight of the body with age for males belonging in the fifteenth generation of the two series of inbreds combined (A, B), and for males in the series of stock controls.

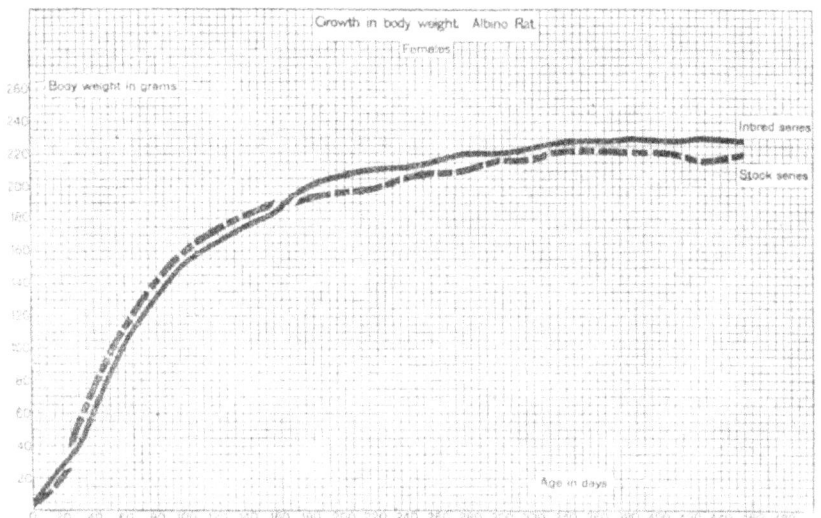

Fig. 14 Graphs showing the increase in the weight of the body with age for females belonging in the fifteenth generation of the two series of inbreds combined (A, B), and for females in the series of stock controls.

35

the stock rats had the advantage of birth in winter. Even under these conditions the males of the fifteenth inbred generation were, on the whole, heavier than the controls after they had attained maturity.

Figure 14 gives graphs showing the weight increase with age for females of the fifteenth inbred generation and for females of the stock control series.

A comparison of the growth graphs for the two groups of females (fig. 14) leads to the conclusion that there was no essential difference in the rate of growth of stock and of inbred females during the early growth period, but that in the adult state the inbred females were slightly heavier at any given age than the females in the control series.

3. VARIABILITY IN THE BODY WEIGHTS OF INBRED RATS

For the purpose of ascertaining the extent of variability in the body weights of the rats in the two inbred series, their coefficients of variation for the body weights at different ages were computed, together with the probable errors. Only records for animals belonging in the seventh to the fifteenth inbred generations were used for this purpose. No attempt was made to find the extent of variability in the body weights of the animals in the first six inbred generations, since only one weight for each animal was recorded and the age period covered by the data at hand was comparatively short.

Table 15 gives the coefficients of variation for the body weights at different ages, with their probable error, for the individuals in the seventh to the fifteenth generations of each of the two inbred series, and for the animals in the two series combined. Grouped data were used in making the calculations for the thirteen and for the thirty day periods, as only the average body weight of the individuals of each sex was recorded in the weighings of the various litters at these ages; for all other ages the individual data were employed.

Comparing the corresponding coefficients for the males and for the females it is found that in each series the female rats

TABLE 15

Showing the coefficients of variation, with their probable error, for the body weights at different ages of the two series of inbred rats (seventh to the fifteenth generation inclusive)

AGE	SERIES A		SERIES B		COMBINED SERIES (AB)	
	Males	Females	Males	Females	Males	Females
days						
13	8.0±0.39	11.8±0.64	11.7±0.74	11.2±0.56	13.2±0.51	12.4±0.45
30	16.6±0.80	16.2±0.88	14.8±0.93	17.4±0.86	16.2±0.62	18.3±0.67
60	19.9±0.96	20.1±1.17	12.8±0.88	14.2±0.70	18.2±0.69	17.7±0.65
90	13.8±0.66	15.6±0.93	11.6±0.73	11.6±0.67	13.8±0.53	13.9±0.58
120	12.9±0.63	10.7±0.64	11.9±0.76	7.6±0.42	12.9±0.51	9.2±0.37
151	11.4±0.56	8.8±0.52	9.2±0.60	7.7±0.45	12.0±0.48	8.9±0.37
182	12.1±0.60	8.0±0.48	11.0±0.72	8.3±0.47	10.2±0.41	8.2±0.34
212	11.4±0.59	9.2±0.57	11.1±0.79	7.7±0.47	12.2±0.51	8.5±0.37
243	11.9±0.62	8.2±0.52	11.4±0.86	8.5±0.55	11.6±0.50	8.1±0.37
273	12.6±0.69	9.0±0.59	11.6±0.97	10.0±0.69	12.3±0.57	9.3±0.45
304	10.7±0.61	8.6±0.64	12.7±1.14	9.5±0.72	11.9±0.57	9.2±0.48
334	9.9±0.60	8.2±0.68	7.9±0.88	9.1±0.85	10.9±0.58	8.5±0.51
365	11.0±0.68	9.0±0.72	6.4±0.74	8.1±0.72	10.5±0.58	8.6±0.51
395	10.6±0.71	7.7±0.66	4.3±0.62	9.2±0.91	9.9±0.60	8.6±0.56
425	11.9±0.85	7.9±0.82	3.0±0.64	10.9±1.19	9.8±0.67	10.9±0.87
455	14.3±1.36	9.6±1.03	3.4±0.98	11.5±1.52	13.7±1.23	11.0±0.91
Average......	12.4±0.70	10.5±0.71	9.7±0.81	10.1±0.73	12.4±0.56	10.7±0.52

were somewhat more variable in body weight than the males during the early stages of development up to sixty days of age, and that beyond this point the males were the more variable. In each series, also, the maximum variability in body weight came at the same age for both sexes: but this maximum was at the sixty day period in the A series of inbreds and at the thirty day period in the B series. The coefficients indicate, moreover, that there was a pronounced tendency in each sex ·for the variability to diminish after the period of rapid growth was ended. Guinea-pigs show a similar decrease in variability in body weight with advancing age, as was noted by Minot ('91).

The average coefficient for the male groups of the A series, taking all ages together, was about two points higher than the corresponding coefficient for the females of this series. This difference is nearly three times the probable error, and is, therefore, large enough to signify that in this series the range of variability in the body weight of the males was greater than that of the females.

In the B series the cofficients for the males were, as a rule, larger than those for the females up to the 334 day period. Beyond this age the coefficients for the females were the larger. This latter relation is not a normal one, and it can be attributed in this instance to the fact that the number of records that were available for use in calculating the coefficients for the older males was very small: in this series only seventeen out of a total of fifty-seven males lived to be one year old.

At all age periods, except at 13 and at 304 days, the coefficients for the males of the A series of inbreds were higher than the corresponding ones for the males of the B series, although the difference in some cases was less than the probable error. Between the average coefficients for the two male groups there was a difference of 2.7 points, but the importance of this difference is greatly lessened by the fact that the coefficients for the older males in the B series were abnormally low. The evidence, on the whole, would seem to indicate that there was little difference in the range of variability in the body weights of the males in the two inbred series, since a comparison of the average coefficients for the two male groups up to 334 days, taking all ages together, shows a difference of only 1.1 points in favor of the individuals in the A series.

At ten of the sixteen age periods shown in table 15 the coefficients of variation for the females of the B series of inbreds exceeded those for the females of the A series, but in many cases the differences were less than the probable error and therefore they can have no significance. The average coefficients for the two groups differed by only 0.04 points, so it is evident that the range of variability in body weights was practically the same for the females of the two inbred series.

A comparison of the coefficients for the males with those for the females in the combined series (A, B) shows that in the majority of cases the male coefficients were much the higher.

At the thirty day period only was the coefficient for the female group significantly greater than the corresponding one for the male group. The greater variability in the body weights of the females at this age is doubtless correlated with the fact that during early postnatal life female rats are growing more rapidly than the males. From the evidence at hand it appears that inbred males are more variable in body weight than inbred females. Jackson ('13) and King ('15 a) have already noted that in groups of stock albinos the males tend to be more variable in body weight than the females at corresponding age periods.

As growth records were taken for only a comparatively small number of animals in each generation of the two inbred series, it did not seem advisable to calculate the coefficients of variation for each generation separately, since Pearson has shown that with numbers less than twenty-five the empirical standard deviation is usually too small. The combined records for the individuals of the two series (A, B), divided into three groups as shown in table 10, were used in calculating the coefficients of variation for the body weights of the generation groups as given in table 16.

As table 16 shows, corresponding coefficients for the three generation groups varied considerably in some cases, but there was a decided tendency in both sexes for all the coefficients to become smaller as the inbred generation advanced. The difference between the average coefficients for the males and for the females in successive generation groups was about two points in each instance. It appears, therefore, that variability in body weight diminished at a fairly uniform rate from one inbred generation to the next. The average decrease for each generation was comparatively slight, amounting to less than one per cent, and it was about the same for the two sexes, although in all generation groups the males tended to be somewhat more variable in body weight than the females.

The question arises as to how the variability in the body weight of the inbred rats compares with that in stock animals in which there is no inbreeding. Obviously in this instance one should compare inbred and stock animals taken from the same strain and reared under similar environmental conditions, since,

TABLE 16

Showing the coefficients of variation, with their probable error, for the body weights at different ages of inbred rats of the two series combined (A, B) separated into three groups according to the generation to which the individuals belonged

AGE	MALES			FEMALES		
	Generations 7-9	Generations 10-12	Generations 13-15	Generations 7-9	Generations 10-12	Generations 13-15
days						
13	13.0 ± 1.06	11.2 ± 0.70	9.4 ± 0.46	14.9 ± 1.09	10.6 ± 0.65	14.4 ± 0.84
30	22.7 ± 1.85	18.4 ± 0.53	11.8 ± 0.69	21.7 ± 1.58	14.8 ± 0.92	15.5 ± 0.91
60	22.9 ± 1.88	21.1 ± 1.46	14.8 ± 0.88	18.8 ± 1.36	16.4 ± 1.32	14.0 ± 0.82
90	15.5 ± 1.26	13.2 ± 0.76	12.4 ± 0.74	18.6 ± 1.52	13.0 ± 0.98	11.4 ± 0.72
120	15.5 ± 1.32	11.8 ± 0.76	11.7 ± 0.70	11.4 ± 0.99	7.5 ± 0.53	9.2 ± 0.56
151	15.1 ± 1.25	11.8 ± 0.76	9.4 ± 0.57	12.2 ± 1.17	7.3 ± 0.50	8.5 ± 0.52
182	13.6 ± 1.13	10.8 ± 0.74	8.6 ± 0.52	10.1 ± 0.87	8.0 ± 0.56	7.1 ± 0.45
212	12.9 ± 1.15	11.1 ± 0.82	8.8 ± 0.55	9.2 ± 0.84	10.3 ± 0.88	6.8 ± 0.41
243	10.8 ± 0.97	11.9 ± 0.88	8.2 ± 0.53	9.5 ± 0.92	9.1 ± 0.80	7.3 ± 0.46
273	12.3 ± 1.17	10.8 ± 0.89	8.7 ± 0.59	11.4 ± 1.21	10.5 ± 1.00	7.6 ± 0.49
304	10.7 ± 1.13	11.3 ± 0.96	9.9 ± 0.69	10.0 ± 1.32	12.1 ± 1.29	7.3 ± 0.49
334	7.8 ± 0.80	9.7 ± 1.03	9.3 ± 0.69	8.6 ± 1.23	11.2 ± 1.38	6.9 ± 0.54
365	6.1 ± 0.68	9.5 ± 0.94	8.6 ± 0.69	9.7 ± 1.64	10.8 ± 1.25	6.5 ± 0.49
395	9.8 ± 1.20	8.5 ± 0.96	6.4 ± 0.56	10.6 ± 2.06	11.3 ± 1.50	5.5 ± 0.44
425	10.6 ± 1.51	8.2 ± 1.01	7.4 ± 0.73	12.3 ± 3.40	10.3 ± 1.64	5.9 ± 0.51
455	10.5 ± 1.57	5.0 ± 0.61	6.0 ± 0.94		7.5 ± 1.26	7.9 ± 0.79
Average......	13.1 ± 1.24	11.5 ± 0.86	9.5 ± 0.66	12.6 ± 1.41	10.7 ± 1.02	8.9 ± 0.59

as several investigators have shown, albino rats from diverse strains that are reared in various ways show pronounced differences in their rate of growth, and presumably there is a corresponding difference in the variability of their body weights at different ages.

Table 17 gives the coefficients of variation for body weights at various ages of fifty male and of fifty female rats reared as controls for the present inbred series. As already stated these animals were a selected group taken from the same strain as the inbred rats and reared under similar environmental conditions. The coefficients of variation for stock rats given in table 17 are reproduced from tables 4, 5 and 6 of a previous publication (King, '15 a).

TABLE 17

Showing the coefficients of variation with their probable error for the body weights at different ages of male and female rats belonging to three series: (1) eight litters of the fifteenth generation of inbreds (series A, B); (2) thirteen litters of selected stock; (3) a single litter of selected stock

AGE	MALES			FEMALES		
	(1) Inbreds 21 rats from 8 litters (15th generation)	(2) Stock 50 rats from 13 litters	(3) Stock 9 rats from 1 litter	(1) Inbreds 27 rats from 8 litters (15th generation)	(2) Stock 50 rats from 13 litters	(3) Stock 7 rats from 1 litter
days						
13	10.4±1.08	11.8±0.79		9.4±0.86	11.4±0.76	
30	10.5±1.08	10.2±0.68		9.3±0.86	11.0±0.74	
60	14.2±1.47	17.0±1.14	9.6±1.52	11.9±1.09	15.7±1.05	6.4±1.15
90	12.9±1.33	14.8±0.99	7.9±1.25	13.2±1.37	12.5±0.95	9.2±1.78
120	11.8±1.23	13.4±0.90	6.5±1.03	8.8±0.86	10.3±0.75	6.5±1.16
151	9.8±1.01	13.3±0.89	5.1±0.80	6.2±0.58	10.4±0.73	8.9±1.59
182	9.1±0.94	14.2±1.22	7.1±1.12	6.7±0.67	12.3±0.90	6.6±1.40
212	8.9±1.00	14.0±0.96	6.8±1.14	5.1±0.48	12.4±0.91	4.9±1.04
243	8.2±0.98	13.9±0.99	10.1±1.69	6.4±0.64	12.6±0.91	7.7±1.38
273	7.5±0.93	13.4±0.99	9.5±1.59	5.4±0.58	11.5±0.89	8.5±1.53
304	5.8±0.72	14.0±1.11	10.0±1.68	3.6±0.40	10.3±0.79	7.4±1.32
334	6.1±0.78	13.7±1.13	8.6±1.44	3.5±0.42	10.8±0.87	6.3±1.22
365	4.0±0.55	13.0±1.16	10.0±1.68	5.2±0.61	10.7±0.91	5.3±1.22
395	4.5±0.67	12.6±1.22	9.8±1.65	4.4±0.57	11.5±0.98	6.5±1.68
425	4.9±0.83	13.4±1.32	13.3±2.23	5.2±0.69	10.9±0.94	5.4±1.15
455	1.6±0.39	13.6±1.67	13.7±2.66	7.5±1.13	8.9±0.99	6.0±1.16
Average......	8.1±0.99	13.5±1.07	9.1±1.53	6.9±0.73	11.4±0.88	6.8±1.34
Average omitting first two records....	7.8			6.6		

A comparison of the coefficients for the males of the combined series (table 15) with the corresponding coefficients for stock males (table 17) shows that inbred males tended to be somewhat more variable in body weight than stock males up to sixty days of age, which is the period of maximum variability for both groups. From this age on, however, stock males were apparently more variable than inbred males. The average coefficient for stock males, taking all ages together, was 13.5, while that for the inbred males was 12.4. Since the difference between these coefficients is only slightly greater than the probable error, it is evident that close inbreeding did not decrease the variability of the entire male population more than about 8 per cent.

When the coefficients for the inbred females of the combined series (table 15) are compared with those for stock females (table 17) it is found that in this sex, also, inbred animals were more variable in body weight than stock animals during the early growth stages. After the age of ninety days, however, stock females tended to be slightly more variable in body weight than inbred females. The average coefficients for the two groups differ by less than one point, so it appears that the inbred females had practically the same range of variability in body weight as the stock females.

In a study of the variability in body weight of the albino rat, Jackson ('13) found that: "variability in body weight is lowest at birth (the coefficient being about 12) and is not much higher at seven days (16). It appears highest at three weeks (28), and at later periods varies from 19 to 21. The average coefficient, taking all ages together, is 19." In Jackson's series the maximum variability in body weight comes at an earlier period than it does in either the inbred series or in the stock controls, and his coefficients are much higher for all ages. The fact that Jackson used data obtained from rats that represented "for the most part a random sample of the general population at each age" undoubtedly accounts for the rather wide range in our results, although a difference in the strain of albinos used and in the environmental conditions to which the rats were subjected may have contributed to the result.

In order to determine whether the animals in the later inbred generations were any less variable in body weight than those in the general inbred population, the coefficients of variation for the body weights at different ages of individuals in eight litters of the fifteenth inbred generation were determined and are given in table 17. All of these coefficients were decidedly lower than the corresponding ones for the individuals of the combined (A, B) series of inbreds (table 15). The difference between the average coefficients was quite large, amounting to 4.3 points for the males and to 3.8 points for the female groups. This result indicates that the animals of the fifteenth inbred generation were about 35 per cent less variable in body weight than the animals in the general inbred population.

When the coefficients of variation for the individuals in the fifteenth inbred generation are compared with those for the stock controls (table 17) the results are equally striking and significant, At only one age period, i.e., thirty days, was the coefficient of variation for the inbred males slightly in excess of that for the stock males; at all other ages the coefficients for the stock males were much the larger. The average coefficient for the stock males was 13.5, while that for the inbred males was only 8.1. It appears, therefore, that the variability in the body weights of the males of the fifteenth inbred generation was, on the average, about 40 per cent less than that in stock males.

Judging from the size of the corresponding coefficients (table 17) females belonging to the fifteenth inbred generation were more variable in body weight at ninety days of age than the females of the stock controls, at all other ages the stock females were the more variable. The difference of 4.5 between the average coefficients for the two groups indicates that the females of the fifteenth inbred generations were about 40 per cent less variable in body weight than stock females.

From a study of fraternal variability in the albino rat, Jackson concludes that "in general the variation in body weight within a given litter of albino rats is probable less than half that of the general population of the same age under similar environment." In the series of stock animals reared as controls for the present inbred series it was shown (King, '15 a) that "the range of variability within the litter is about 70 per cent that of the general population in the case of the males, while for the females it is about 55 per cent."

By comparing corresponding coefficients, as given in table 17, it is possible to determine whether variability in the body weight of the animals in the fifteenth inbred generation was greater or less than fraternal variability in stock animals. The evidence as given indicates that the inbred males were more variable in body weight than the males in a single stock litter up to 212 days of age, but that at all subsequent ages stock males were very much more variable. Omitting from the inbred series the coefficients for the first two age periods, since there were no corresponding coefficients for the males of the stock litter, the average coefficient for the males in the fifteenth inbred generation, taking all ages together, was 7.8, while the average coefficient for the stock males was 9.1. The difference between the coefficients was not very great, but it seems large enough to signify that the variability in the body weights of the males in the fifteenth inbred generation was somewhat less than fraternal variability in the stock controls.

The relative variability of inbred and stock females was slightly different from that found in the male groups. Females of the fifteenth inbred generation were, as a rule, more variable in body weight than females in a single stock litter up to 120 days of age, beyond this age stock females seemed to be the more variable. The difference of 0.2 points between the average coefficients for the two groups, omitting the findings for the first two weighings of the inbred group, was in favor of the stock litter. This difference, however, is much too small to have any meaning, and it is evident that the variability in the body weights of the females of the fifteenth inbred generation was about the same as fraternal variability in the stock controls.

4. DISCUSSION OF RESULTS

This study of the growth in body weight of albino rats belonging to fifteen generations produced by brother and sister matings has shown that the closest form of inbreeding possible in mammals does not necessarily produce animals that are below the normal body size, as Crampe ('83) and Ritzema-Bos ('93; '94) have maintained. During the early part of these experiments all of the evil effects that are said to follow from close inbreeding were obtained, but it was shown conclusively that they were not due to inbreeding but solely to malnutrition. There is a possibility, therefore, that many of the bad results obtained in other inbreeding experiments with rodents may have been produced, in part at least, by unfavorable environmental or nutritive conditions. The published records of the former work give no details regarding the manner in which the experiments were conducted, consequently there is no way of determining to what extent external conditions were responsible for the outcome. Both Crampe and Ritzema-Bos worked with hybrids, which frequently exhibit a tendency to sterility as others have noted, and the animals were inbred promiscuously for the most part. Even the most favorable environmental conditions could not be expected to keep such animals up to normal standards for body size or for fertility.

In the present series of experiments improper feeding through four successive generations did not permanently impair the growth power of the individuals, which responded at once to the stimulus of a well balanced diet. The rats in the fifth and those in the sixth inbred generations grew much more vigorously than their forefathers, and many of them attained an adult size equal to that which is normal for the albino rat.

The maximum effect of the stimulus given to the growth impulse by adequate nourishment did not seem to be reached until the seventh generation when some of the animals were larger than any albino rats as yet recorded. In the two following generations the average body weight of both males and females at various age periods decreased slightly, but they were still far above the norms for stock animals and higher than the averages for the individuals in the eleventh and succeeding generations. From the tenth to the fifteenth generations there was no very marked change in the average body weights of the rats at corresponding age periods, but the body weights tended to decrease slightly as the inbreeding advanced.

It seems probable that the exceptionally vigorous growth of the rats in the seventh to the ninth inbred generations was wholly, or in great part, a response of the organisms to very favorable nutritive conditions following a period of partial starvation.

Hatai ('07), Osborne and Mendel ('14; '15; '16) and Stewart ('16) have shown that the growth of the albino rat can be inhibited for varying periods, either by starvation or by improper food, and that there is at once a resumption of growth when a return is made to a normal diet; the animals eventually reaching the size of the controls or even surpassing them in body weight at corresponding ages.

Although, as a rule, adult rats increase in body weight very slowly and may even remain at practically the same body weight for several successive months, this slowing up of the growth process is apparently not due to an exhaustion of the growth capacity. The extensive experiments of Osborne and Mendel show that the capacity to grow can be retained and exhibited at periods far beyond the age at which growth ordinarily ceases, and their work points to the conclusion that in the rat "the capacity to grow is only lost by the exercise of this fundamental property of animal organisms." What is true for the individual may also be true for the race. The capacity to grow is seemingly so essential a part of the organism that this power is retained through several successive generations in which it is not exercised to its full extent. In this series of experiments, as far, as is known, not a single individual of the many hundreds that were reared in the first four generations attained a body size that equaled the norm for the adult albino rat. Yet even after this long period of time the growth impulse in all individuals at once responded to the stimulus of adequate nutrition, and only two generations were required to effect a return to normal body size.

That favorable nutritive conditions had produced a parallel modification of the soma and of the germplasm might be a satisfactory explanation for the appearance of the exceptionally large individuals in the seventh to the ninth inbred generations were it not for the fact that this increase in the body size of the individuals was temporary, lasting through these generations only. It seems more probable that favorable nutritive conditions, following a period of semi-starvation, greatly increased metabolic activity and so stimulated the growth impulse that the animals attained an unusually large size. After the maximum effect of the stimulus had passed there was a gradual decline to more normal conditions of metabolism and a corresponding decrease in the average size of the individuals. I see no reason to assume that the hereditary factors concerned in growth were influenced either by malnutrition during the early part of the experiment, or by favorable nutrition in the later generations.

Crampe ('83) and Hoskins ('16) have noted that the growth of albino rats is influenced to a considerable extent by the time of year in which the animals are born. Rats born in the winter months are larger at a given age, live longer and are more vigorous than those born in the summer or autumn. The superiority of the winter-born rats has been most marked in the various breeding experiments that I have been carrying on for several years with different strains of rats. It is not improbable that the rats of the seventh inbred generation owed some part of their vigorous growth to the fact that they were born in the most favorable season of the year, the early winter months. The environmental agency here concerned in stimulating growth is either temperature or humidity, possibly both. A moderate degree of cold is apparently more conducive to rapid and vigorous growth in rats than is heat: the reverse is true for the mouse, according to the investigations of Sumner ('09; '15). Extreme temperature, either heat or cold, has a very unfavorable effect on the rat, making the animals exceeding susceptible to the rat scourge, pneumonia, which invariably proves fatal to an animal of any age.

In these experiments, as already stated, there was a very careful selection of breeding stock from the seventh generation on. Small, weak, inferior animals were eliminated before reaching maturity, and only the largest and most vigorous animals were allowed to perpetuate their kind. By this rigid selection it was possible to keep the animals up to a high standard for body size and to make the strain apparently immune to the injurious effects that would probably have followed from random matings. In other inbreeding experiments where selection of breeding animals was made on the basis of size and vigor there was no

marked deterioration in the stock.   Castle ('16 a) inbred rats
within narrow lines of selection for seventeen successive generations and was able to maintain races 'with fair vigor and
fecundity.'   Inbreeding experiments with Drosophila, carried on
for many generations by Castle et al. ('06) and by Moenkhaus
('11), have shown that in this form, also, races of large size and
vigor can be maintained under the closest inbreeding by simply
selecting the most vigorous parents for breeding.

Undoubtedly various species of plants and animals react differently under inbreeding.   In tobacco, inbreeding is "beneficial
and offers an effective means of maintaining desirable characteristics in the established varieties" (Shamel, '05); while in maize
inbreeding leads to a considerable loss in vegetative vigor but
not to degeneration (East and Hayes, '11; '12).   In swine, according to Darwin ('76), close inbreeding invariably leads to
sterility and to a considerable loss in body size after only a few
generations, although this has recently been denied by Gentry
('05).   From available evidence inbreeding seems to be very injurious to dogs (Darwin, '76; B, '06), to pigeons (Fabre-Domengue,
'98) and to goats (Ewart, '10).   On the other hand, it is chiefly
through inbreeding, with selection, that the thoroughbred types
of cattle, of sheep and of horses have been developed and fixed,
and there is no evidence of degeneration in the descendants
of deer and of rabbits that have been inbred in isolated communities for many years (Ewart, '10).   If pure stock that ful-
fills standard requirements as to size, vigor and fertility is used
for the investigation, and only vigorous, sound animals are
allowed to breed, there is, theoretically, no ground for believing
that continued inbreeding will cause either loss of vigor or a
decrease in body size.   In these experiments with the rat, the
bad effects of inbreeding per se, as far as they might manifest
themselves by a decrease in the body size of the individuals, have
apparently been entirely prevented through the use of a strain of
animals that seemingly had no inherent defects and by a careful
selection of breeding stock.   Even after twenty-eight generations of continued brother and sister matings the inbred animals
have not deteriorated in any way, and they are still superior

in body size to stock animals reared under similar environmental conditions.

Brother and sister matings automatically tend to reduce heterozygosis, and by the time that the animals have reached the fifteenth generation they are 96.277 per cent homozygous (Fish, '14). Such inbred individuals, according to Pearl ('13), can by no chance possess more than 0.006 of 1 per cent of the different lines of ancestral descent which are theoretically possible. In the present experiments selection was also a factor that tended to decrease heterozygosis, since the mating of only the largest and most vigorous pair of individuals in a litter presumably brought together gametes of like genetic constitution, in the majority of cases, and thus aided in increasing the proportion of homozygotes in the progeny population. In spite of the very high degree of homozygosis which they had attained, the animals of the fifteenth inbred generation showed a considerable amount of variability in body weight at different ages (table 17). How much of this variability was due to genetic factors for body size and how much was purely the result of environmental and nutritive action is not known, since there is, as yet, no means of determining to what extent environmental agencies can influence body growth. Nutritive conditions alone can greatly alter body size in the rat, as is shown by the experiments of Osborne and Mendel ('16). Temperature and humidity likewise act upon body growth in rodents (Sumner, '09; '15), and housing conditions are known to materially change their body size. With all of these environmental agencies that are known to affect body size made as uniform as possible under existing laboratory conditions, the variability in the body weights of the inbred rats continued to decrease at a small, but fairly uniform, rate in both males and females from the seventh to the fifteenth generation. This indicates that the individuals were becoming more homozygous with respect to the factors that determine body size. It is evident, however, that even after fifteen generations of continued brother and sister matings the strain was far from 'pure' in the sense in which this term is used by Johannsen ('09) and his followers, since the data for body growth in the individuals of later generations, which will be given in a subsequent publication, show a still further decrease in the variability of body weights as the inbreeding advanced.

As the coefficients of variability were not determined for any individuals in the first six inbred generations, the extent of variability in the body weights of the inbred series as a whole is not known. A comparison of the average coefficients for body weight as given in table 15 and in table 17 show that the animals in the seventh to the fifteenth inbred generations were, as a group, about 7 per cent less variable in body weight than the animals in the control series. This difference would doubtless be greater if the inbred rats had not undergone unusual changes in body weight due to altered conditions of nutrition.

On the basis of the calculations made by Fish ('14) regarding the amount of homozygosis in different generations of animals produced by brother and sister matings, the group of rats comprising the seventh to the ninth inbred generation was, on the average, about 83 per cent homozygous. These animals, as table 16 shows, had a range of variability in body weight that was greater than that in any other generation group. It is interesting to note that, although the body weights of this group of inbreds greatly exceeded the norms for stock animals of like age, their variability was not correspondingly increased, since the average coefficients for the group are about the same as those for the stock controls (table 17). The group comprising the individuals of the tenth to the twelfth inbred generations was, according to Fish's table, about 91 per cent homozygous, while the last group was, on the average, 95 per cent homozygous. Since body weight probably depends to a considerable extent on "the presence or absence of definite genetic factors segregating from one another in gametogenesis on lines with which we are already familiar" (Punnet and Bailey, '14), one might expect to find a definite correlation between the amount of homozygosis and the variability in body weight in animals obtained from brother and sister matings. On referring to the average coefficients for the body weights of the various generation groups, as given in table 16, it is found that for both males and females there was, in every case, a difference of about two points between the coefficients for successive generation groups. In these inbred rats, therefore, variability in body weight did not decrease proportionately as the homozygosis of the individuals increased. The results are complicated by the fact that part, at least, of the variability in body weight that was measured by the coefficients of variability was due to environmental action and cannot be distinguished from the variability due to genetic growth factors. When the data for a number of later inbred generations have been examined it may then be possible to ascertain the probable extent of variability due to environment and to correlate the amount of homozygosis in the individuals with the extent of their variability in body weight.

The records, as given above, show unquestionably that the variability in body weights of the individuals decreased as the inbreeding advanced. The results, therefore, do not accord with Walton's ('15) theory that continued inbreeding tends to increase rather than to diminish variability. In an able criticism of this theory Castle ('16 c) says: "It is difficult to understand how on any theory of heredity inbreeding could be expected to increase variability within a single inbred line. . . . . On a Mendelian theory it would be expected that inbreeding, brother with sister, for a large number of generations would result in the production of a number of homyzygous lines, each of which by itself would be entirely devoid of variability, except that due to environmental agencies." The results so far obtained in this inbreeding experiment with the rat are in harmony with Castle's view, though in such a complex organism as the rat it is not probable that any degree of inbreeding will produce lines that are "entirely devoid of variability, except that due to environmental agencies."

In discussing the effects of close inbreeding on Drosophila, Castle et al. ('06) state: "These experiments show no appreciable effect of inbreeding. In every case the brood reared under the best and most uniform conditions has the highest average number of teeth (on the sex comb), irrespective of whether or not inbred. The same may be said of variation in size. Inbreeding has diminished neither the average size nor the variability in size." The reactions of the rat to close inbreeding are slightly different from those of Drosophila. The closest form of inbreeding possible, continued for many generations, has not caused a diminution in the average body weight of inbred rats at any age. On the contrary, through the selection of only the largest and most vigorous animals for breeding, inbred animals, especially the males, are superior in body size to the best stock animals reared under similar environmental conditions. In the rat variability in body weight decreases after inbreeding, and in the fifteenth inbred generation the variability was no greater than fraternal variability in the body weights of stock albinos.

The more general bearing of the results of these inbreeding experiments will be discussed in a following paper dealing with the fertility and constitutional vigor of inbred rats.

SUMMARY

1. The present paper gives an analysis of the data for the increase in the weight of the body with age for 333 male and for 306 female albino rats. These animals belonged to two series (A and B), both descended from the same ancestral stock, that were inbred, brother and sister only, for fifteen successive generations.

2. The animals of the first six generations suffered from malnutrition, and their body weights were much smaller than the norms for stock animals of like age. Many of these animals had defective teeth and the majority of the females were sterile. When nutritive conditions were improved the animals quickly regained their normal body size and the tendency to sterility and to malformation was checked.

3. The general course of the growth in body weight of inbred rats is similar to that of stock animals as determined by the investigations of Donaldson, Jackson and others.

4. In both inbred series the average body weights of the males was greater than that of the females at every age at which weighings were taken. The males and females of the B series

were somewhat heavier at all ages than the animals of the A series.

5. The males belonging in the seventh to the ninth generations of the two inbred series greatly exceeded the males of the first six inbred generations in body size, and they were, as a rule, much heavier, at all ages, than the males of the subsequent generations (fig. 6).

6. The females belonging in the first six generations of the two inbred series were considerably smaller, at all ages, than the females of the subsequent generations. Adult females of the seventh to the ninth inbred generations were slightly larger than the females of the later generations (fig. 7).

7. The unusual size of the individuals in the seventh to the ninth inbred generations was probably due, in great part, to the stimulus given to the growth impulse by favorable nutritive conditions following a prolonged period of semi-starvation.

8. Inbred males belonging in the seventh to the fifteenth generation inclusive were heavier at all ages than stock albinos. In the adult state the inbred males were, on the average, 18 per cent heavier than the general run of stock albinos and about 12 per cent heavier than males from a selected stock series reared under the same environmental condition (fig. 11).

9. Inbred females were, as a rule, slightly heavier at any given age than the females of the control series, but the difference between the two groups was much less than in the case of the males. At the 365 day period the average body weight of the inbred females was 3.7 per cent greater than that of the stock females (fig. 12).

10. Inbred males were more variable in body weight than inbred females, the maximum variability for both sexes coming before the animals were two months old. These results agree with the findings for stock albinos as determined by the investigations of Jackson and of King.

11. The males of the A series of inbreds were slightly more variable in body weight than the males of the B series, but the females of the two series showed practically the same variability in body weight at corresponding age periods (table 15).

12. In both series the variability in the body weights of males and of females decreased as the inbred generation advanced. The average decrease was about 2 per cent for a group of three generations (table 16).

13. Inbred males were more variable than stock males up to sixty days of age. After this time stock males showed greater variability in body weight at all ages (tables 15, 17).

14. Inbred females were more variable in body weight than stock females up to ninety days of age. After reaching maturity stock females tended to be slightly more variable in body weight than inbred females (tables 15, 17).

15. Males and females of the fifteenth inbred generation were about 35 per cent less variable in body weight than the animals of the general inbred population, and about 40 per cent less variable than the animals in the series of stock controls (table 17).

16. The variability in the body weights of the males of the fifteenth inbred generation was somewhat less than fraternal variability in the males of the stock controls. In the corresponding female groups variability in body weight was practically the same at all ages (table 17).

STUDIES ON INBREEDING

II. THE EFFECTS OF INBREEDING ON THE FERTILITY AND ON THE CONSTITUTIONAL VIGOR OF THE ALBINO RAT

HELEN DEAN KING

The Wistar Institute of Anatomy and Biology

TWO CHARTS

The present paper gives data showing the fertility and the constitutional vigor in a strain of albino rats that was inbred, litter brother and sister, for twenty-five successive generations. Details regarding the manner in which these experiments were conducted and data for the growth and variability in the body weight of inbred rats have already been published (King, '18a).

1. THE FERTILITY OF INBRED RATS

As shown by a number of recent investigations (Pearson et al., '99; Rommell and Phillips, '06; Pearl, '12, a, and Wentworth, '16), fertility is undoubtedly a racial character that is transmitted by inheritance, although it is influenced to a considerable extent by a variety of extraneous factors. The mode of inheritance of fertility in the rat is not discussed in the present instance, since the effects of inbreeding on fertility is the chief subject under consideration.

Throughout this paper the word 'fertility' is used as defined by Pearl and Surface ('02) to designate: "The total actual reproductive capacity of pairs of organisms, male and female, as expressed by their ability when mated together to produce (i.e., bring to birth) individual offspring." According to this view, fertility depends upon and includes fecundity as well as a great number of other factors. As Pearl and Surface state: "Clearly it is fertility rather than fecundity which is measured in statistics of birth of mammals."

The inbred strain of rats was composed of two series, A and B, both derived from the same ancestral stock. In every generation of each series the females that were used for breeding were paired twice with a brother from the same litter, thus producing the strictly 'inbred' litters that alone furnished the breeding stock in the following generation. These same females were then paired twice with an unrelated Albino male taken from the general stock colony. For convenience, litters with the latter parentage are here designated as 'half-inbred' litters.

The early generations of these inbred animals suffered severely from malnutrition, due to improper feeding. Nutritive conditions were improved after the fourth generation, and the animals quickly regained their normal size and fertility. At no stage of the investigation was any attempt made to influence the productiveness of the animals, other than by keeping them under environmental and nutritive conditions that were as uniform and as favorable as it was possible to make them.

A. Litter size

The normal fertility of any race can properly be estimated only from the total number of offspring produced by many females during the entire period of their reproductive activity. The fertility in the inbred strain of rats cannot be measured by this standard, unfortunately, since the plan of the experiment called for only four litters from each breeding female, and after this number was obtained the females were usually discarded. According to Crampe ('84), the Albino female has, on the average, only three or four litters. On this basis the litter data obtained for the inbred series shows the total productiveness of the greater proportion of the females that were used for breeding. Crampe's estimate for litter production is, I believe, too low, since the breeding history of a considerable number of stock Albinos, recently obtained, shows that the females had an average of 5.3 litters each. Records for the inbred series undoubtedly cover the most productive period in the life of the females, and if the fertility of the strain was impaired to any extent by inbreeding it is probable that all of the litters cast would have been smaller than normal.

The number and average size of the litters produced in each of the first twenty-five generations of the A series of inbreds are given in table 1. Similar data for the litter production in the B series of inbreds are shown in table 2.

These tables are inserted chiefly for reference, but a comparison between corresponding data indicates clearly that the fertility of the animals in one inbred series was about the same as that in the other series. The summary of the data for the two series shows that the 1752 litters in the A series contained an average of 7.5 young, while the 1656 litters in the B series had an average of 7.4 young. This close agreement in the records for two such large groups of animals is doubtless due to the fact that all of the inbred rats were descended from the same ancestral stock and that individuals in corresponding generations of the two series were reared simultaneously under similar environmental conditions.

The data in table 1 and in table 2 have been combined in table 3, which thus shows the number and average size of the litters produced in the first twenty-five generations of the inbred strain. The data given comprise the records for 3408 litters containing 25,452 individuals.

To facilitate the discussion of the effects of inbreeding on fertility the data given in table 1 to table 3 were combined by generation groups. There were relatively few individuals in the first six generations of inbreds and their data were united to form the first group, since the character of the experiment was changed at this point. Data for subsequent generations were divided into five groups, each of which, with the exception of the last, comprised the records for four successive generations. Such a division of the data was, of course, purely arbitrary, but it seemed the most satisfactory arrangement possible. A group of four generations covers approximately the litter production for one year, and as the number and size of the litters vary considerably at different times of the year, this grouping assured a uniform distribution of the seasonal variations in litter size among all of the various groups.

Litter data for the A series of inbreds, arranged according to generation groups, are given in table 4.

TABLE 1

Showing the number and the average size of the litters produced in each of the first twenty-five generations of the A series of inbred rats

LITTER SERIES

GENERATION	First litters (inbred)			Second litters (inbred)			Third litters (half-inbred)			Fourth litters (half-inbred)			SUMMARY		
	Number of litters	Number of individuals	Average number of young per litter	Number of litters	Number of individuals	Average number of young per litter	Number of litters	Number of individuals	Average number of young per litter	Number of litters	Number of individuals	Average number of young per litter	Total number of litters	Total number of individuals	Average number of young per litter
1	1	7	7.0	1	7	7.0	1	6	6.0	1	9	9.0	4	29	7.2
2	3¹	20	6.7	4	27	6.5	3	19	6.3	3	14	4.6	13	80	6.1
3	7	35	5.0	5	26	5.2	5	27	5.4	5	21	4.2	22	109	5.0
4	13	58	4.4	12	77	6.4	8	52	6.5	6	46	7.6	39	233	6.0
5	18	117	6.5	18	130	7.2	15	111	7.4	10	58	5.8	61	416	6.8
6	15	94	6.2	15	102	6.8	14	86	6.1	11	63	5.7	55	345	6.2
7	16	101	6.3	16	129	8.0	15	109	7.2	9	61	6.7	56	400	7.1
8	17	122	7.1	17	142	8.3	15	100	6.6	8	56	7.0	57	420	7.3
9	17	105	6.2	17	123	7.2	16	121	7.5	12	76	6.3	62	425	6.8
10	20	131	6.5	20	156	7.8	20	161	8.0	17	137	8.0	77	585	7.6
11	21	144	6.8	21	170	8.1	20	164	8.0	18	135	7.5	80	613	7.6
12	20	145	7.2	20	154	7.7	19	152	8.0	17	140	8.2	76	591	7.7
13	22	160	7.2	22	180	8.1	21	156	7.4	20	157	7.8	85	653	7.6
14	21	139	6.6	21	188	8.9	21	182	8.1	18	136	7.5	81	645	7.9
15	23	171	7.4	23	193	8.3	21	173	8.2	17	136	8.0	84	673	8.0
16	21	143	6.8	21	165	7.9	18	121	6.7	10	62	6.2	70	491	7.1
17	27	210	7.7	27	244	9.0	25	213	8.5	23	178	7.7	102	845	8.2
18	23	168	7.3	23	197	8.5	19	164	8.6	15	110	7.3	80	639	7.9
19	23	147	6.3	23	178	7.7	22	172	7.8	15	118	7.8	83	615	7.4
20	27	193	7.1	27	224	8.3	23	149	6.5	17	130	7.6	94	696	7.4
21	27	199	7.4	27	217	8.0	26	223	8.5	22	187	8.5	102	826	8.1
22	27	203	7.5	27	234	8.6	24	218	9.0	17	113	6.6	95	768	8.0
23	25	186	7.4	25	183	7.3	22	161	7.4	19	142	7.4	91	672	7.3
24	25	184	7.3	25	200	8.0	22	177	8.0	21	140	6.6	93	701	7.5
25	26	173	6.6	26	203	7.8	23	166	7.2	15	104	6.9	90	646	7.1
1–25	485	3,355	6.9	483	3,849	7.9	438	3,383	7.7	346	2,529	7.3	1,752	13,116	7.5

¹ One female destroyed her first litter.

TABLE 2

Showing the number and the average size of the litters produced in each of the first twenty-five generations of the B series of inbred rats

LITTER SERIES

GENERATION	First litters (inbred)			Second litters (inbred)			Third litters (half-inbred)			Fourth litters (half-inbred)			SUMMARY		
	Number of litters	Number of individuals	Average number of young per litter	Number of litters	Number of individuals	Average number of young per litter	Number of litters	Number of individuals	Average number of young per litter	Number of litters	Number of individuals	Average number of young per litter	Total number of litters	Total number of individuals	Average number of young per litter
1	1	5	5.0	2	19	9.5	2	17	8.5	1	9	9.0	1	5	5.0
2	2	12	6.0	5	26	5.2	2	9	4.5				7	57	8.1
3	7	36	5.1	8	69	8.6	7	56	8.0	3	18	6.0	14	71	5.0
4	11	66	6.0	20	158	7.1	18	150	8.3	13	95	7.3	29	209	7.2
5	20	133	6.5	15	118	7.8	14	102	7.2	8	74	9.2	71	536	7.5
6	15	106	7.0	15	105	7.0	15	104	6.9	10	78	7.8	52	400	7.7
7	15	79	5.8	15	129	8.6	13	94	7.2	3	21	7.0	55	366	6.6
8	15	108	7.6	20	183	9.0	17	127	7.4	10	49	4.9	46	352	7.6
9	20	126	6.2	17	145	8.5	16	107	6.6	14	93	6.6	67	485	7.2
10	17	104	6.1	19	128	6.7	18	138	7.6	18	149	8.2	64	449	7.0
11	19	112	5.9	20	175	8.7	19	154	8.1	12	68	5.6	74	527	7.1
12	20	139	6.9	21	168	8.0	19	150	7.9	13	104	8.0	71	536	7.5
13	21	154	7.3	21	135	6.4	20	161	8.0	18	132	7.3	74	576	7.7
14	21	142	6.7	20	138	6.9	20	173	8.6	17	140	8.2	80	570	7.1
15	20	131	6.5	24	184	7.6	22	190	8.6	19	132	6.9	77	582	7.5
16	24	158	6.5	22	170	7.7	21	173	8.2	17	113	6.6	89	664	7.4
17	22	158	7.1	23	180	7.8	20	171	8.5	20	149	7.4	82	614	7.4
18	23	168	7.3	24	221	9.2	22	195	8.8	13	98	7.4	86	668	7.7
19	24	186	7.7	26	179	6.8	21	157	7.4	11	88	8.0	83	698	8.4
20	26	174	6.7	24	189	7.8	18	138	7.6	16	112	7.0	84	598	7.1
21	24	176	7.3	26	210	8.0	24	211	8.8	19	139	7.3	82	615	7.5
22	26	174	6.7	22	195	8.8	20	145	7.2	15	126	8.4	95	734	7.8
23	22	181	8.2	27	189	7.0	26	175	6.7	17	114	6.7	79	647	8.1
24	27	207	7.6	26	193	7.4	25	179	7.1	20	136	6.8	97	685	7.0
25	26	184	7.0										97	692	7.1
1–25	468	3,219	6.9	462	3,606	7.8	419	3,276	7.8	307	2,235	7.2	1,656	12,336	7.4

TABLE 3

Showing the number and average size of the litters produced in each of the first twenty-five generations of the two inbred series (A,B): a combination of the data in table 1 and in table 2

GENERATIONS	First litters (inbred)			Second litters (inbred)			Third litters (half-inbred)			Fourth litters (half-inbred)			SUMMARY		
	Number of litters	Number of individuals	Average number of young per litter	Number of litters	Number of individuals	Average number of young per litter	Number of litters	Number of individuals	Average number of young per litter	Number of litters	Number of individuals	Average number of young per litter	Total number of litters	Total number of individuals	Average number of young per litter
1	2	18	6.0	1	7	7.0	1	6	6.0	1	9	9.0	5	34	6.8
2	5	32	6.4	6	46	7.6	5	36	7.2	4	23	5.7	20	137	6.8
3	14	71	5.1	10	52	5.2	7	36	5.1	5	21	4.2	36	180	5.0
4	24	124	5.1	20	146	7.3	15	108	7.2	9	64	7.1	68	442	6.5
5	38	250	6.6	38	288	7.6	33	261	7.9	23	153	6.6	132	952	7.2
6	30	200	6.6	30	220	7.3	28	188	6.7	19	137	7.2	107	745	6.9
7	31	180	5.8	31	234	7.5	30	213	7.1	19	139	7.3	111	766	6.9
8	32	230	7.2	32	271	8.5	28	194	6.9	11	77	7.0	103	772	7.5
9	37	231	6.2	37	306	8.3	33	248	7.5	22	125	5.7	129	910	7.0
10	37	235	6.3	37	301	8.1	36	268	7.4	31	230	7.4	141	1,034	7.3
11	40	256	6.4	40	298	7.4	38	302	8.0	36	284	7.9	154	1,140	7.4
12	40	284	7.1	40	329	8.2	38	306	8.1	29	208	7.2	147	1,127	7.6
13	43	314	7.3	43	348	8.0	40	306	7.6	33	261	7.9	159	1,229	7.7
14	42	281	6.7	42	323	7.7	41	343	8.4	36	268	7.5	161	1,215	7.6
15	43	302	7.0	43	331	7.7	41	346	8.4	34	276	8.1	161	1,255	7.8
16	45	301	6.7	45	349	7.7	40	311	7.6	29	194	6.0	159	1,155	7.2
17	49	368	7.5	49	414	8.4	46	386	8.4	40	291	7.2	184	1,459	7.9
18	46	336	7.3	46	377	8.2	39	335	8.5	35	259	7.4	166	1,307	7.8
19	47	333	7.1	47	399	8.5	44	367	8.3	28	214	7.6	166	1,313	7.9
20	53	367	6.9	53	403	7.6	44	306	7.0	28	218	7.7	178	1,294	7.2
21	51	375	7.3	51	406	8.0	44	361	8.2	38	299	7.9	184	1,441	7.8
22	53	377	7.0	53	444	8.3	48	429	8.9	36	252	7.0	190	1,502	7.9
23	47	367	7.8	47	378	8.0	42	306	7.3	34	268	7.9	170	1,319	7.7
24	52	391	7.5	52	389	7.5	48	352	7.3	38	254	6.7	190	1,386	7.7
25	52	357	6.8	52	396	7.6	48	345	7.2	35	240	6.9	187	1,338	7.2
1–25	963	6,574	6.9	945	7,455	7.9	857	6,650	7.8	653	4,764	7.3	3,408	25,452	7.5

TABLE 4

Showing for various generation groups the number and average size of the litters produced in the A series of inbred rats

LITTER SERIES

GENER-ATIONS	First litters (inbred)			Second litters (inbred)			Third litters (half-inbred)			Fourth litters (half-inbred)			SUMMARY		
	Number of litters	Number of individuals	Average number of young per litter	Number of litters	Number of individuals	Average number of young per litter	Number of litters	Number of individuals	Average number of young per litter	Number of litters	Number of individuals	Average number of young per litter	Total Number of litters	Total Number of individuals	Average number of young per litter
1-6	57	331	5.8	55	369	6.7	46	301	6.5	36	211	5.9	194	1,212	6.2
7-10	70	459	6.5	70	550	7.8	66	491	7.4	46	330	7.1	252	1,830	7.2
11-14	84	588	7.0	84	692	8.2	81	654	8.0	73	568	7.8	322	2,502	7.7
15-18	94	692	7.3	94	799	8.5	83	671	8.0	65	486	7.5	336	2,648	7.8
19-22	104	742	7.1	104	853	8.2	95	762	8.0	71	548	7.7	374	2,905	7.7
23-25	76	543	7.1	76	586	7.7	67	504	7.5	55	386	7.0	274	2,019	7.4
1-25	485	3,355	6.9	483	3,849	7.9	438	3,383	7.7	346	2,529	7.3	1,752	13,116	7.5

An examination of table 4 shows that all of the litters produced in the first generation group of the A series were smaller, on the average, than corresponding litters in the later generation groups. The relatively low fertility of the animals in the early generations was not due to inbreeding, but to the fact that these individuals suffered from malnutrition. As soon as the nutritive conditions were improved there was at once an increase in the number and in the size of the litters produced, as the data for the fifth and for the sixth inbred generations show (table 1).

As indicated in the last column of table 4, the groups comprising the tenth to the twenty-fifth generations of the A series showed, as a whole, comparatively little variation in the average size of the litters. The maximum average size (7.8) came in the group including the fifteenth to the eighteenth generations. This maximum was, however, only 0.1 greater than the average litter size for the preceding and for the following group, and therefore it can have little, if any, significance.

Litter data for various generation groups in the B series of inbreds are shown in table 5.

As the average size of the litters produced in the first generation group was greater than that in the second group (table 5), it might appear that the fertility of the breeding females in the B series was not lessened by malnutrition. In the beginning of these experiments many more females of the B series than of the A series were completely sterile, but the females of the B series that did breed were the more productive. Malnutrition, in this instance, was a selective agent that helped to eliminate the tendency to sterility in the B series by preventing the breeding of any except the most fertile females.

In the B series the maximum average size of the litters was found in the group comprising the nineteenth to the twenty-second generations, but, as was the case in the A series, this maximum was not great enough to be considered significant.

Litter data given in table 4 and in table 5 have been combined in table 6.

The data for each of the two inbred series, as well as that given in table 6, shows that in all generation groups the first litter cast

TABLE 5

Showing for various generation groups the number and average size of the litters produced in the B series of inbred rats

GENER-ATIONS	LITTER SERIES												SUMMARY		
	First litters (inbred)			Second litters (inbred)			Third litters (half-inbred)			Fourth litters half-inbred					
	Number of litters	Number of individuals	Average number of young per litter	Number of litters	Number of individuals	Average number of young per litter	Number of litters	Number of individuals	Average number of young per litter	Number of litters	Number of individuals	Average number of young per litter	Total number of litters	Total number of individuals	Average number of young per litter
1- 6	56	358	6.3	50	390	7.8	43	334	7.7	25	196	7.8	174	1,278	7.3
7-10	67	417	6.2	67	562	8.4	61	432	7.0	37	241	6.5	232	1,652	7.1
11-14	81	547	6.7	81	606	7.4	76	603	7.9	61	453	7.4	299	2,209	7.4
15-18	89	615	6.9	89	672	7.5	83	707	8.5	73	534	7.3	334	2,528	7.5
19-22	100	710	7.1	100	799	8.0	85	701	8.2	59	435	7.3	344	2,645	7.6
23-25	75	572	7.6	75	577	7.7	71	499	8.4	52	376	7.2	273	2,024	7.4
1-25	468	3,219	6.9	462	3,606	7.8	419	3,276	7.8	307	2,235	7.2	1,656	12,336	7.4

TABLE 6

Showing for various generation groups the number and average size of the litters produced in the two inbred series (A,B): a combination of the data in table 4 and in table 5

	LITTER SERIES												SUMMARY		
GENERATIONS	First litters (inbred)			Second litters (inbred)			Third litters (half-inbred)			Fourth litters (half-inbred)					
	Number of litters	Number of individuals	Average number of young per litter	Number of litters	Number of individuals	Average number of young per litter	Number of litters	Number of individuals	Average number of young per litter	Number of litters	Number of individuals	Average number of young per litter	Total Number of litters	Total Number of individuals	Average number of young per litter
1-6	113	689	6.1	105	759	7.2	89	635	7.0	61	407	6.9	368	24,90	6.8
7-10	137	876	6.4	137	1,112	8.1	127	923	7.2	83	571	6.9	484	3,482	7.2
11-14	165	1,135	6.8	165	1,298	7.9	157	1,257	8.0	134	1,021	7.6	621	4,711	7.5
15-18	183	1,307	7.1	183	1,471	8.0	166	1,378	8.3	138	1,020	7.4	670	5,176	7.7
19-22	204	1,452	7.1	204	1,652	8.1	180	1,463	8.1	130	983	7.6	718	5,550	7.7
23-25	151	1,115	7.4	151	1,163	7.7	138	1,003	7.3	107	762	7.1	547	4,043	7.4
1-25	953	6,574	6.9	945	7,455	7.9	857	6,659	7.8	653	4,764	7.3	3,408	25,452	7.5

was the smallest in the litter series, as a rule; the second litter was the largest; the third and fourth litters were intermediate in size between the first and the second; a similar relation in the size of litters has been found, also, in two groups of stock Albinos. In both inbred series the first two litters cast in each generation were the offspring of brother and sister matings; the third and fourth litters were produced by the mating of an inbred female with an unrelated stock male. .In the first of the stock series noted (King and Stotsenburg, '15, table 7) all of the litters were produced by the pairing of unrelated stock animals; in the second stock series, for which data are given in table 7 of the present paper, all of the litters obtained were the offspring of brother

TABLE 7

Showing the average size of each of the first four litters produced by a series of stock Albino females

LITTER SERIES	NUMBER OF LITTERS	NUMBER OF INDIVIDUALS	AVERAGE NUMBER OF YOUNG PER LITTER
1	116	717	6.2
2	116	843	7.3
3	103	671	6.5
4	89	587	6.6
	424	2818	6.7

and sister matings. Since in all three groups the average size of the litters in the litter series varied in a similar way, it is evident the litter size does not depend at all on the relatedness or the unrelatedness of the parents, but chiefly on the age of the female. Young females tend to be somewhat less prolific than older ones, as Crampe ('83) noted. The litters reach their maximum size when the females are about five months old, but the number of young does not decrease appreciably in the various litters cast until the females have passed the height of their reproductive power at about seven months of age (King, '16b).

As shown in the first paper of this series (King, '18), the rats in the seventh to the ninth inbred generations were considerably heavier, at any given age, than the individuals belonging to subsequent generations. The cause for this unusually vigorous growth was attributed to a stimulation of the growth processes produced by adequate nutrition following a period of semi-starvation. During this period the productiveness of the females was increased considerably, since the average size of the litters in the group comprising the seventh to the tenth generations was 0.4 greater than the average for the previous generation group (table 6). The period of maximum fertility in the inbred series did not, however, coincide with the period of maximum growth in body weight, but came at a much later time (fifteenth to the twenty-second generations), when the litters contained 7.7 young, on the average. The fact that the average size of the litters in the last three generations of the inbred series was slightly lower than the maximum can be attributed to a change in diet made necessary by the economic conditions of the present time. This diet does not seem to be quite as favorable to growth and fertility as was the more varied diet used until the beginning of last year.

The graph in figure 1, showing the average size of the litters produced in the various generations of the inbred strain, was constructed from the data in the last column of table 3.

Starting at the point of 6.8, the graph in figure 1 drops at the third generation to 5.0, the lowest point in its course. From this point it rises slowly, and after the fifth generation tends to be a fairly horizontal line, since it never falls below 6.9 nor does it rise above 7.9. At more or less regular intervals the graph drops slightly below the normal level. The most pronounced depression is at the point of the third generation; a second drop comes at the ninth generation; other depressions of about the same depth are found at the point of the sixteenth, the twentieth, and the twenty-fourth generations. As the last three depressions in the graph occur at intervals of four generations, it is evident that they were not due to a chance variation in the data, but that they must express periodic changes in the reproductive cycle of the females that tended to reduce the number of young born. In whatever way this reduction was effected, whether by a lessening of fecundity or by limiting the number of embryos that were capable of normal development, the cause for it, I believe, lay in the seasonal changes in temperature which always have a marked effect on the

physical condition of the animals.   During the summer months
rats suffer severely from excessive humidity and from high temperature, since their mechanism for heat regulation, under these
conditions, is inadequate.   At this season their sexual activity
is at its lowest point, and the litters that are produced tend to be
relatively small.   Severe cold checks reproduction, but litters
born under these conditions are usually of normal size and the

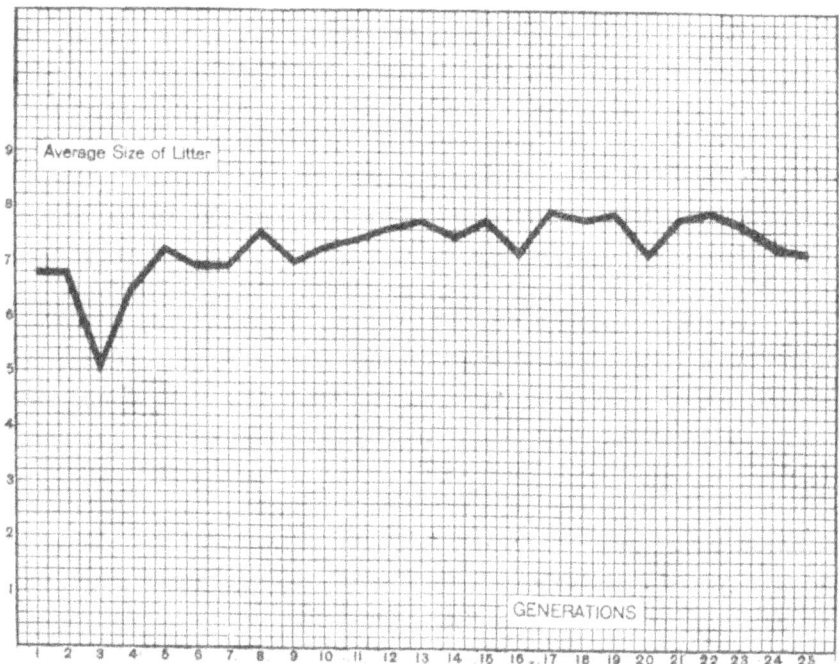

Fig. 1   Graph showing the average size of the litters produced in the various
generations of the inbred strain (data in table 3).

individuals strong and vigorous.   At about every fourth generation the majority of litters produced in the inbred strain were
born at the most unfavorable season of the year, the summer and
early fall.   In this generation, as the records show, the litters
were smaller, as a rule, than the litters in the preceding and in the
following generations.   This decrease in size was sufficient to
account for all of the depressions in the graph in figure 1, except
the first one, which was doubtless due to the fact that the maxi-

mum effect of malnutrition in lowering the fertility of the females was reached at the third generations.

Cyclic changes in productiveness were noted by Castle et al. ('06) in an inbred strain of Drosophila, in which, for three successive years, there was a gradual rise in fertility followed by an abrupt decline. These changes in productiveness were likewise ascribed to the variations in temperature at different seasons of the year.

The data given in table 1 to table 6 and the graph in figure 1 show clearly that, despite all theories to the contrary, it is possible to maintain a high degree of fertility in a mammal for at least twenty-five generations of the closest possible form of inbreeding, by a careful selection of breeding stock and by keeping the animals under environmental conditions that are favorable for their growth and reproduction.

While, in general, the size of the litter varies according to the age of the mother, individual females differ greatly regarding the number of offspring that they produce in any litter of the litter series. Sisters from the same litter, mated to the same male, will show marked variations in their fertility at the same age. One female may never have a litter that contains more than five young; the other may always throw litters in which there are nine or more young. Some females, regardless of their age, tend to cast the same number of offspring in every litter. One female so noted had ten young in each of her four litters. Marked individual differences in fertility are also found among female guinea-pigs, according to Minot ('92).

The average number of young in a litter of albino rats is 6.3, according to the data for 394 litters collected by Crampe ('84); Cuénot ('99) found an average of 8.5 young in the 30 litters that he examined. Records for 1089 litters of stock Albinos born in the Wistar Institute animal colony during the years 1911 to 1914 give 7.0 as the average number of young per litter (King and Stotsenburg, '15). When this last series of data was collected it was not realized that litter size in the rat depends to such a marked degree upon the age of the mother, and that in this species the maximum fertility comes at a relatively early age, as it does in the human race (Powys, '05) and also in poultry (Pearl, '17). Most of the litters recorded were cast by young females that had not reached the height of their reproductive power; such litters tend to be larger than those cast after this time (King, '16b). Data for litters cast by females of unknown age, however extensive they may be, cannot, therefore, properly be used to furnish the norm for litter size in the albino rat.

In order to obtain standards for litter size with which the data in the inbred strain might justly be compared, the complete breeding history of a considerable number of stock Albino females was recorded during the past three years. Data for the first four litters produced by 116 females belonging to this group are given in table 7. All of the stock rats from which these litters were obtained were reared under the same environmental conditions as the inbred strain.

In table 7, as has already been noted, the litters of the series bear the same size relation to each other as that found in the litter series of the inbred rats. The first litter was the smallest, averaging 6.2 young; the second litter, with an average of 7.3 members, was the largest of the series; while the third and fourth litters were somewhat smaller than the second. The entire series of 424 litters gave an average of 6.7 young per litter. This average is 0.3 less than that in the random collection of stock litters previously recorded (7.0), and 0.4 more than the norm as given by Crampe (6.3), so it is probably a fair standard for litter size in any similar series of Albino litters. There is no reason to believe that the stock females from which the litters recorded in table 7 were obtained were, as a group, inferior in reproductive power to other stock females, and presumably their fertility at any given age is fairly representative of that in the general run of stock Albinos.

Each litter of the stock series, shown in table 7, contained a smaller average number of young than the corresponding litter in either of the two inbred series when the data were arranged according to generation groups (tables 4 and 5), and, omitting the records for the first five generations where the fertility was lessened by malnutrition, there was not a single generation in either of the inbred series where the average size of the first four litters was as low as that in the stock series (tables 1 and 2). In the inbred series as a whole, the average size of the litters was 0.8 greater than that in the stock series. Even if the previous finding of 7.0 be taken as the norm for litter size in the rat, the difference between the average size of the litters in the inbred strain and the norm chosen is 0.5. This difference is great enough to preclude the possibility that it was due to chance, and it cannot be attributed to the differential action of environment, since stock and inbred rats were constantly under the same environmental conditions. According to these findings, fertility in the inbred strain of Albinos, in as far as it may be judged by the size of the first four litters cast by a large number of females, was greater than the fertility in stock Albinos that were not inbred.

B. Frequencies of litter size

According to the several series of observations that have been recorded, there is a wide range in the size of the litters cast by Albino females. Kolazy ('71) reports litters containing from five to seventeen young, although Crampe ('84) states that he never found even fourteen young in a litter of albino rats. Litter size varied from four to twelve in the series of Albinos studied by Kirkham and Burr (15); while in the litters recorded by King and Stotsenburg ('15) the range in size was from two to fourteen.

Data for litter frequencies in the two series of inbred rats are shown in table 8.

TABLE 8

Showing the frequencies of litter size in the two series of inbred rats

	SIZE OF LITTER																
	1	2	3	4	5	6	7	8	9	10	11	12	13	14	15	16	17
A	1	40	59	103	193	182	288	268	258	181	99	45	27	5	2	0	1
B	0	35	72	102	168	206	263	260	208	152	110	49	22	6	3	0	0
	1	75	131	205	361	388	551	528	466	333	209	94	49	11	5	0	1

In the A series of inbreds the range in litter size was from one to seventeen. The litter of one was cast by a female of the nineteenth generation that was suffering from pneumonia and had to be killed three days after parturition. This is undoubtedly a case where the physical condition of the female prevented the normal development of all of the embryos except one; the other embryos probably became atrophic and were absorbed. The litter containing seventeen members occurred in the fifteenth generation. All of the individuals were born alive, but they were all very small, weighing not more than three grams each: the average weight of the albino rat at birth is about four grams (King, '15b).

In the B series, as table 8 shows, the range of variation in litter size was not as great as that in the A series: no litters smaller than two or larger than fifteen were obtained. In both series litters containing seven young were the most frequent, while those with eight young were only slightly less in number.

Figure 2 shows graphs for litter frequencies in the two series of inbreds that were constructed from the data given in table 8.

In figure 2 each graph rises quickly to the modal point at seven, falls slowly at first and then rapidly. The drop in graph A at the point of 6 has apparently no significance, since there is no similar drop in the B graph. Each graph is a simple frequency curve with one modal point, and is exactly the sort of graph that one would expect to obtain from the data for litter frequencies in a large series of animals belonging to a pure race.

C. Puberty

Under normal conditions puberty tends to appear at approximately the same age in the different individuals of a given race, but the time of its appearance is seemingly more dependent on the growth changes incident to age than it is on age itself.

In the albino rat both males and females attain sexual maturity when they are about two months of age (Donaldson, '15), but the environmental and nutritive factors that hasten or retard growth have considerable influence on the reproductive activity of the

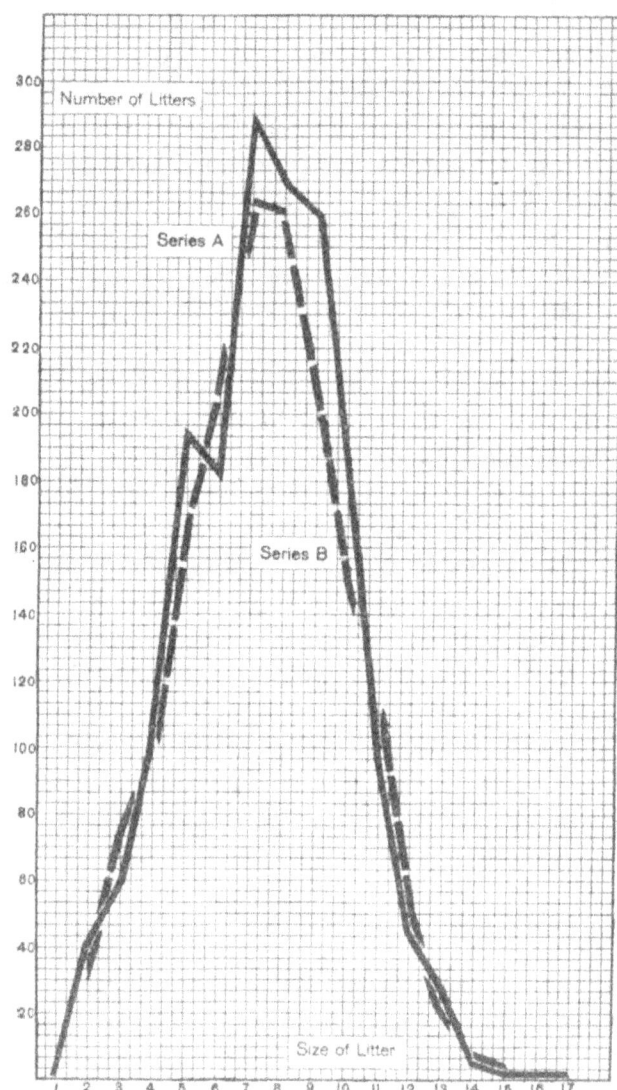

Fig. 2 Graphs showing the frequencies of litter size in the two series of inbred rats (data in table 8).

individuals. If young rats are fed exclusively on a meat diet, puberty is considerably delayed (Watson, '06); the same effect is produced by underfeeding (Osborne, Mendel and Ferry, '17). According to my observations, the time of year in which the animals are born affects their subsequent growth and also the time of their maturing. Rats born in the winter and early spring grow rapidly, and usually breed at about three months of age; those born in the summer and autumn grow more slowly and comparatively few of the females cast litters before they are four months old, many not breeding until spring, which is the season of the most pronounced sexual activity for the rat. Convincing evidence that age alone does not determine the beginning or the end of the reproductive life of the rat is given by Osborne and Mendel ('15, '17), who found that Albino females, stunted at an early age by underfeeding, were completely sterile until they were properly nourished, when they grew rapidly, attained a normal size, and were able to breed long after the age at which the menopause usually appears.

According to Darwin ('75) and others, favorable environment tends to delay sexual maturity, though not necessarily to decrease fertility. Since these inbred rats were reared, for the most part, under environmental conditions that seemed well adapted to their needs, and since they lacked the stimulus to reproductive vigor which is said to come from outcrossing, it might be expected that they would tend to mature much later than stock Albinos which were not inbred.

Table 9 shows the approximate age at which the breeding females belonging to various generation groups of the two inbred series cast their first litter.

The records for the first generation group, given in table 9, confirm Osborne and Mendel's findings that underfeeding tends to retard sexual maturity, since they show that about one-half of the breeding females in this group did not cast their first litter until they were four months old. In subsequent generations, when the animals were adequately nourished, they began breeding at a much earlier age. Under the conditions of this experiment, inbreeding seemingly hastened the onset of puberty, for in both series, as the inbreeding advanced, there was a marked tendency for relatively more of the females to breed at the earliest possible age. About 30 per cent of the breeding females in the eighteenth to the twenty-fourth generation group of each series threw their first litter at or before the age of ninety days; only a small proportion of them failed to breed before reaching the age of four months.

As a whole, the females of the A series of inbreds tended to mature slightly earlier than the females of the B series, but the

TABLE 9

Showing the approximate age at which breeding females in various generation groups of the two inbred series (A and B) cast their first litters

GENERATION GROUPS TO WHICH BREEDING MALES BELONGED	SERIES A				SERIES B				SUMMARY (A, B)			
	Total number of breeding females	Per cent females breeding before 90 days of age	Per cent of females breeding between 90 and 120 days of age	Per cent females breeding after 120 days of age	Total number of breeding females	Per cent females breeding before 90 days of age	Per cent females breeding between 90 and 120 days of age	Per cent females breeding after 120 days of age	Total number of breeding females	Per cent females breeding before 90 days of age	Per cent females breeding between 90 and 120 days of age	Per cent females breeding after 120 days of age
1– 5	58	12.0	36.2	51.8	56	14.3	41.1	44.6	114	13.1	38.7	48.2
6– 9	70	21.4	68.6	10.0	67	7.4	74.7	17.9	137	14.6	71.6	13.8
10–13	84	25.0	71.4	3.6	81	13.6	74.1	12.3	165	19.3	72.8	7.9
14–17	94	28.7	62.8	8.5	89	25.8	62.9	11.3	183	27.3	62.9	9.8
18–21	104	36.5	62.6	0.9	100	31.0	64.0	5.0	204	33.8	63.3	2.9
22–24	76	31.5	60.6	7.9	75	20.0	69.4	10.6	151	25.8	65.0	9.2
1–24	486	27.1	61.6	11.3	468	19.8	65.3	14.9	954	23.5	63.4	13.1

difference between the two series was not great, and corresponding records were in nearly as close agreement as were those for litter size.

The youngest breeding female in the inbred strain was a member of the A series of inbreds, and she was eighty days old when she cast her first litter of five young. As the gestation period in the albino rat is about twenty-two days (Donaldson, '15), this female must have conceived when she was two months old. Kirkham and Burr ('15) state that one of their Albino females gave birth to a litter when she was only seventy-seven days old; while Lantz ('10) reports a case in which an albino rat was said to have produced a litter at the age of fifty-six days. This last case is certainly a remarkable one, and its parallel has not been found among the 50,000 rats bred in our colony.

The last section of table 9 shows that, after the tenth generation, there was no marked change in the proportion of females that bred at three and at four months of age, respectively. Nearly 24 per cent of the total number of females used for breeding cast their first litter by the time they were three months old; over 60 per cent of them bred for the first time when they were between ninety and one hundred and twenty days of age; while about 13 per cent did not breed until after they were four months old. The latter group was made up, for the most part, of females that were born in the summer or autumn.

Although Düsing ('84) states that inbred animals tend to mature very early, I do not think that inbreeding alone was responsible for the fact that relatively more of the females in the later than in the earlier generations of these inbred rats bred at three months of age. In these experiments, when two or more females of a litter were reared as possible breeding stock, the first female that became pregnant was the one taken to continue the line, provided she fulfilled all requirements as to size and vigor. Thus the manner in which breeding females were selected preserved those individuals that tended to breed at an early age, and this tendency to early maturity, if heritable, must have been retained in the stock and intensified to some extent through continued brother and sister matings. Inbreeding, aided by selection, would·thus seem to be the factor involved in producing a strain of rats in which the females attained sexual maturity at a relatively early age.

D. Sterility

Sterility occurs normally in the Albino, as in other strains of rats, and therefore it might be expected to appear at times in any strain, regardless of whether the animals were inbred or outbred. Crampe ('84) states that of 221 Albino females which he selected for breeding forty-six, or 20.8 per cent, were sterile.

Out of 124 stock Albino females reared in our own colony during the past three years and intended for breeding purposes thirty-two, or 28.8 per cent, were completely sterile, while about 10 per cent of those that did breed cast only one or two litters. Unfortunately, no records have been kept that give information regarding the exact proportion of sterile females in the first six generations of the inbred series. The number was relatively very large, and must have included at least one-half of the total number of females that lived to be six months old. Sterility in these females was, for the most part, the result of poor nutrition, and it disappeared as soon as the nutritive conditions were improved. Loeb ('17) has shown that in the guinea-pig "under-feeding prevents maturation of the follicles and thus causes sterility which lasts as long as the effect of the underfeeding is present in the ovary." In the guinea-pig, as in the rat, adequate nutrition reëstablishes normal conditions in the ovary and sterility almost entirely disappears.

In Drosophila, according to Castle et al. ('06), low productiveness (sterility) is directly transmitted by inheritance and is amenable in selection. In the rat, sterility seems to depend not entirely on genetic factors, but to a marked extent upon conditions, such as malnutrition and disease, that act unfavorably upon reproduction. In the present experiments, by selecting for breeding only the most vigorous individuals (which it seems were also the most fertile), sterility in as far as it may depend on genetic factors would seem to have been practically eliminated from the strain, and it has not reappeared even after twenty-eight generations of brother and sister matings.

Of the 954 inbred females that were used for breeding during the course of these experiments, 653, or 68.5 per cent, cast four litters each, and many of them, kept for body-weight records, produced several other litters which were not recorded. Of the females that did not cast the required four litters, the great majority died from pneumonia, or were killed because they showed unmistakable evidence of illness. A few of the females stopped breeding after producing one or two litters, although they were apparently in good physical condition and were paired for several months with males that were known to be fertile. A postmortem examination of the reproductive organs from several of these semi-sterile females showed, in every instance, an inflamed condition of the ovaries or of the uterus which would render reproduction impossible. Barrenness in these cases was doubtless due to disease and not to any inherent tendency to sterility. A similar diseased condition of the reproductive organs has been found to be responsible for the partial sterility of stock Albinos.

2. THE CONSTITUTIONAL VIGOR OF INBRED RATS

The best criterion by which to gauge the so-called 'constitutional vigor' of any animal is undoubtedly its power of reproduction, since that is of the utmost importance for the continuation of the race. There are, however, other important tests for vigor that can be applied, such as the rate and extent of growth, agility, mental alertness, resistance to disease, and ability to live to an advanced age. According to Darwin ('78), "the effects of close interbreeding in animals, judging from plants, would be deterioration in general vigor, including fertility, with no necessary loss of excellence of form." This would seem to indicate that, whatever tests were applied, closely inbred animals and plants would show marked inferiority when compared with individuals of the same species that were not inbred. That such a sweeping generalization is not justified is shown by the results of a number of recent inbreeding experiments: the work of Shamel ('05) on tobacco, of Stout ('16) on chicory, and of Hayes and Jones ('17) on tomatoes give no indication that self-fertilization in these plants causes a loss either of vegetative or of reproductive vigor; Gentry's ('05) experiments on swine, and those of Castle et al. ('06) and of Moenkhaus ('11) on Drosophila show that any loss of vigor that might come from inbreeding can be entirely overcome by the proper selection of breeding stock.

The present series of experiments on the rat are the first recorded for any mammal in which brother and sister matings were made continuously for twenty-five successive generations. In the inbreeding experiments with rodents made by Crampe, by Ritzema-Bos, and by von Guaita, matings were made between animals related in various degrees, and they were made as often between parent and offspring as between sibs. Ritzema-Bos states: "Bemerkenswert ist namentlich das Result, dass die Paarung zwischen Geschwistern viel schlechtere Erfolge lieferte als die Paarung zwischen Mutter und Sohn, resp. Vater und Tochter." Presumably, therefore, my inbred strain, in which all breeding females came from litters produced by the matings of sibs only, would show an even greater evidence of deterioration in vigor than did the rats inbred by Crampe and by Ritzema-Bos.

Data already given show that these inbred rats were much more fertile than stock rats reared under the same environmental conditions, so it is evident that their reproductive vigor was not impaired. In their ability to withstand disease inbred rats compared favorably with stock rats. The rat scourge, pneumonia, was quite as prevalent among stock animals as among the inbreds and took its toll of lives as frequently and as quickly in one strain as in the other. Parasitic infection was as common in the stock colony as in the inbred, and severe changes in temperature were followed by just as many deaths among stock animals as occurred in the inbred strain. The rat's power of resistance to disease and to unfavorable environmental conditions did not appear to be lessened by inbreeding under the conditions of these experiments.

Records for the growth in body weight of a considerable number of rats belonging in various generations of the two inbred series show the approximate age at which death occurred in all individuals that did not live to the end of the weighing period, which came when the rats were fifteen months old. As similar records were recently obtained for a series of stock animals, it is possible to compare the relative length of life in the two strains and thus to determine whether inbreeding tends to shorten the life of the individuals, as it might be expected to do if it impaired the general vigor of the animals to any extent.

Table 10 shows the mortality at different ages in such of the A series of inbreds as were used for the determination of the effects of inbreeding on the growth in body weight, given in the first paper of this series. For convenience the data were arranged in generation groups: the last group includes the findings through the twenty-third generation only, as the weight records for animals belonging in the twenty-fourth and in the twenty-fifth generations are not yet completed. As all of the animals reached the age of three months, the first mortality record given is that for animals at six months of age.

On examining the mortality data for the males, as given in table 10, it is found that comparatively few of the animals in any generation group died before the age of six months, and that over 50 per cent of them lived to be more than one year old. A comparison of the corresponding records for the various generation groups shows unmistakably that the animals belonging to

TABLE 10

Showing the mortality at different ages in a group of 236 males and 179 females belonging in the seventh to the twenty-third generations of the A series of inbred rats

GENER-ATION GROUPS	NUM-BER OF MALES	PER CENT MALES LIVING AT VARIOUS AGES				NUMBER OF FEMALES	PER CENT FEMALES LIVING AT VARIOUS AGES			
		6 mos.	9 mos.	12 mos.	15 mos.		6 mos.	9 mos.	12 mos.	15 mos.
7–10	35	91.4	71.4	54.3	45.7	28	92.8	67.8	25.0	10.7
11–14	52	90.3	71.1	57.7	38.4	37	97.3	83.8	56.7	35.1
15–18	60	100.0	75.0	53.3	26.6	47	97.8	72.3	59.5	29.8
19–23	89	98.8	88.7	73.0	46.0	67	98.5	82.1	67.1	46.2
7–23	236	96.2	78.8	61.9	39.4	179	97.2	77.6	56.4	34.1

the later generations tended to be longer lived than did those in the earlier generations.

The mortality data for the females of the A series are much like those for the males, the most noticeable difference being found in the records for the first generation group where only 10 per cent of the females lived to be fifteen months of age. Taking the animals of the A series as a whole, about 4 per cent of them died before they reached the age of six months; 20 per cent did not live to the age of nine months; 50 per cent were dead at the end of one year, and only about 35 per cent lived to be fifteen months old.

Mortality data for individuals belonging to various generation groups of the B series are shown in table 11.

In the earlier generations of the B series the mortality in both males and females was considerably greater than that in the animals belonging to the A series: only 5 per cent of the males lived to be fifteen months old, while not a single female reached this age. For the later generation groups the data for the B series were very similar to those for the A series. As a whole, however, the animals in the A series lived longer than did those in the B series.

The data given in table 10 and in table 11 have been combined in table 12. This table shows also mortality data for 377 stock albino rats reared in The Wistar Institute animal colony during the past three years. Included in the latter series are the

TABLE 11

Showing the mortality at different ages in a group of 151 males and 231 females belonging in the seventh to the twenty-third generations of the B series of inbred rats

GENER-ATION GROUPS	NUM-BER OF MALES	PER CENT MALES LIVING AT VARIOUS AGES				NUMBER OF FEMALES	PER CENT FEMALES LIVING AT VARIOUS AGES			
		6 mos.	9 mos.	12 mos.	15 mos.		6 mos.	9 mos.	12 mos.	15 mos.
7–10	18	94.4	50.0	38.8	5.5	34	79.4	23.5	17.6	
11–14	30	86.6	70.0	26.6	6.6	43	90.7	65.1	34.9	16.2
15–18	43	100.0	69.8	51.1	27.9	64	96.9	76.5	54.6	26.5
19–23	60	100.0	93.3	76.6	56.6	90	100.0	86.6	77.7	53.3
7–23	151	97.2	76.8	54.9	32.4	231	94.3	70.6	54.5	31.1

records, elsewhere published (King ,'15), for fifty males and for fifty females of selected stock that were reared as controls for the inbred strain.

The mortality data for the inbred rats, given in table 12, show that close inbreeding did not tend to shorten, but to lengthen the span of life in both males and females: 50 per cent of the animals belonging to the last group lived to be fifteen months of age, while in none of the other groups did even 30 per cent of the individuals attain this age. It is probable that the relatively high death rate in the animals of the earlier generations was due to the fact that the rats had not regained the vigor that was so greatly impaired in their ancestors by malnutrition.

Donaldson ('06) has assumed that the span of life in man is thirty times that of the rat, and therefore that a rat of three years corresponds to a man of ninety years. Considering the relatively small proportion of men that live to be nonagenarians, one would not expect to find many rats in any colony living to three years of age, yet under the equitable climate of California, Slonaker ('12) succeeded in keeping two of a series of sixteen albino rats beyond this age, and one of them lived for forty-five months, or the equivalent of one hundred and twelve years of human life. At various times during the past five years a number

TABLE 12

Showing the mortality at different ages in a group of 387 males and 410 females belonging in the seventh to the twenty-third generations of the two inbred series (a combination of the data in table 10 and in table 11). Data are also shown for the mortality in a series of stock albino rats comprising 199 males and 178 females

GENERATION GROUPS	NUMBER OF GROUPS	PER CENT MALES LIVING AT VARIOUS AGES				NUMBER OF FEMALES	PER CENT FEMALES LIVING AT VARIOUS AGES			
		6 mos.	9 mos.	12 mos.	15 mos.		6 mos.	9 mos.	12 mos.	15 mos.
7–10	53	92.4	64.1	49.0	32.0	62	85.4	43.5	20.9	4.8
11–14	82	89.0	70.7	47.1	26.8	80	93.7	73.7	45.0	25.0
15–18	103	100.0	72.8	52.4	27.1	111	97.3	74.8	56.7	27.9
19–23	149	99.3	90.6	74.5	50.3	157	99.3	84.9	73.3	50.3
7–23	387	96.3	78.0	59.2	36.7	410	95.6	73.6	55.3	32.0
Stock series..	199	98.9	85.9	63.8	28.1	178	95.5	87.6	70.2	37.6

of inbred and of stock Albinos were kept in our colony in good physical condition until they were about two years old. We have never attempted to keep any rats beyond this age.

Osborne, Mendel and Ferry ('17) state that out of ninety-one albino rats kept under ordinary laboratory conditions during their entire lifetime, "17 (19 per cent) died under one year of age; 48 (53 per cent) died between one and two years of age; and 26 (29 per cent) lived more than two years, the oldest one reaching an age of nearly 34 months. From these figures it is evident that less than a third of the rats in our colony may be expected to live to be more than two years old." In another paper these

authors ('15) state: "Fully half of our stock rats have died before
the age of 600 days." Unfortunately, the mortality data given
by Osborne, et al. are not in a form which makes it possible to
compare them directly with the records for these inbred rats.
It would seem, however, from the results as given, that their
animals tended to live longer than the rats in my inbred strain.
The mortality data for the series of stock Albinos, given in table
12, can be directly compared with those for the inbred rats given
in the same table, since both series of animals were reared under
similar environmental conditions and the records were taken at
the same age intervals. Relatively more of the stock than of the
inbred males were living at six, nine, and twelve months of age,
but only 28 per cent of the stock males lived to be fifteen months
old, while 37 per cent of the inbred males attained this age.
The records for the female groups show that relatively as many
inbred as stock females lived to the age of six months, but that
more of the stock than of the inbreds were living at nine, twelve,
and fifteen months of age. Taken as a whole, therefore, longevity in the inbred strain seemed to be somewhat less than that
in the stock controls.

Some of the inbreeding data for animals which Darwin ('75)
collected were so at variance with his own results on plants that
he was forced to admit that; "manifest evil does not usually
follow from pairing the nearest relations for two, three, or even
four generations." In a long-continued series of inbreeding
experiments, therefore, the deleterious effects of inbreeding
would supposedly be more accentuated in the later than in the
earlier generations. A comparison between the mortality records for stock animals and those for the inbred group comprising
the animals in the nineteenth to the twenty-third generation
should show the effects of inbreeding on longevity much better
than the comparison between the groups as previously made.
Such a procedure is the more justifiable, perhaps, because these
two groups of animals were reared in the colony simultaneously.
While in the two male groups only about 1 per cent of the animals
failed to reach the age of six months, relatively more of the inbred
than of the stock males were living at all other age periods noted:

the final records for the two groups show a difference of 22.2 per cent in favor of the inbred animals. In the female groups the span of life in the inbreds also tended to be longer than that in the controls, but the difference was not quite as marked as in the case of the males: the final records show a difference of only 12.7 per cent.

It appears, from the above comparison of data for stock and inbred rats, that continued inbreeding, under favorable environmental conditions and with the aid of selection, cannot only lessen the tendency to early death caused by malnutrition, but that it can extend the average span of life in the rat considerably beyond that found in the stock controls. Constitutional vigor, as judged by the longevity of the individuals, is therefore not invariably lessened by continued inbreeding.

In table 10 and in table 11 it will be noted that the mortality data for the first generation group indicate that the span of life in the females, particularly in the B series, was much shorter than that in the males. The reason for this 'selective mortality' is not clear, although it may be that the females were not able to throw off the effects of malnutrition quite as readily as were the males. In both inbred series, after the tenth generation, the mortality in the females at any age period was practically the same as that in the corresponding group of males. Data given in table 12 show that stock females tended to live longer than stock males: a reversed relation seemed to hold for the inbred rats. Taking the inbred colony as a whole, I am inclined to the opinion that the females, as a rule, tend to live longer than do the males. More males than females usually die as the result of a sudden, sharp change in temperature, and the impression one gets from working daily with the animals is that the males are far more susceptible to pneumonia than are the females, and that they are sooner attacked by various parasitic pests, such as lice and ear-mites. White ('14) states that in India the bubonic plague is a more fatal disease to male than to female rats, thus indicating that the female is stronger, constitutionally, than the male. These results are in accord with the findings for the human race: census reports and various statistical tables that have been compiled show, as does the investigation of Pearson et al. ('03), that the duration of life in women is longer than it is in men and that women are the less susceptible to disease at all ages.

The various physical defects, so prevalent among Crampe's ('83) inbred rats, were all found among my inbred rats at the beginning of these experiments, but they were due to malnutrition, not to inbreeding, since they entirely disappeared when the animals received proper food. Among the thousands of inbred animals that were reared during the past five years some few, not to exceed a dozen in all, lacked one or both eyeballs. This defect has also appeared, at times, in stock animals. On the average, one in every 10,000 rats born in the stock colony is tailless. This abnormality, as Conrow ('15, '17) has shown, involves the skeletal structure in the entire pelvic region. The inbred colony has contained only one tailless individual as yet. Unfortunately, this rat was destroyed by the mother soon after birth, so it was not carefully examined. Neither of these defects appears to be heritable, and neither can be due to inbreeding, since each has appeared also in a stock that is outbred. No other abnormalities of any kind have appeared in the animals of the inbred strain up to the present time when the individuals of the twenty-eighth generation are approaching maturity. The findings in this series of experiments, therefore, do not give support to Ritzema-Bos' contention that inbreeding tends to cause "eine grössere Prädisposition für Krankheiten und das Entstehen von Missbildungen." When a considerable number of animals belonging to any series exhibits various kinds of malformations, it is safe to assume that either environmental and nutritive conditions are unfavorable to normal development, as in the early part of the present series of experiments, or that there is an inherent weakness in the stock used that is brought out and accentuated by random inbreeding, as seemed to be the case with Crampe's rats.

No data are available for a direct comparison between stock and inbred rats as regards their relative activity at different ages, but several series of experiments have been made in different psychological laboratories in which the behavior of rats from this inbred strain was compared with that of stock controls.

In the inbred rats of the earlier generations the brain and spinal cord were decidedly below the normal weight of these organs in stock animals of like age and body weight. "From the fourth to the tenth generation the relative brain weight remained, on the average, constant at six and one-half per cent less than that of the normal control rats" (Basset, '14). The habit formation in a number of rats that belonged in the sixth and in the seventh inbred generations was tested at Johns Hopkins University by Basset ('14), who found that these animals were inferior to stock rats in their ability to form habits, and that they show less retention of a habit, and were longer in relearning it, than were the controls.

Inbred rats belonging in the twelfth and in the fourteenth generations were sent to Harvard University where Mrs. Yerkes ('16) studied their behavior and compared it with that of stock albino rats obtained from The Wistar Institute colony and from a different source of supply. The general conclusion reached by Mrs. Yerkes was that "inbred rats learned a trifle more slowly than the stock rats, both in the maze and in the discrimination experiments, but that they carried discrimination of lightness and darkness further, and showed the most pronounced difference only in their greater timidity and instability of behavior."

Temperamental differences between stock Albinos and inbreds of the fourteenth and the fifteenth generations were investigated at Harvard by Utsurikawa ('17). The results obtained showed that inbred rats were less active and more savage than the outbred rats, and that they responded more quickly and in greater amount to momentary auditory stimulation than did outbred rats. The two strains were found to differ also in "restlessness or continuity of response." Inbred rats showed the greatest restlessness" in case of momentary and repeated auditory stimulation and less in case of continued stimulation, whereas for the outbred animals the reverse is true." These temperamental differences between inbred and stock rats would seem to indicate that inbred rats are more 'high strung' nervously than are outbred rats. Nervousness is a trait manifested by many thoroughbred animals, and it is particularly characteristic of the racehorse.

The nervousness of the horse is undoubtedly the result of continued selection, since breeders consider that an animal must have this trait highly developed if it is to be a success on the track. If nervousness is a trait that is transmitted by inheritance and amenable to selection it is probably also a trait that would tend to be intensified by close inbreeding, and therefore it might be expected that rats closely inbred for many generations would be somewhat more nervous than outbred stock controls, as Utsurikawa found to be the case.

When the last two series of investigations were completed the animals used were sent to The Wistar Institute where they were killed and carefully examined by Dr. Hatai. It was found, as Mrs. Yerkes states, that the inbred rats had a somewhat greater body length and body weight than the stock rats, and that they showed a brain weight in relation to body length and body weight that was only from 0.002 per cent to 0.006 per cent less than that of stock rats. Since the inbred rats of the sixth and of the seventh generations had a brain weight about six and one-half per cent less than the normal (Basset, '14), it would appear, from Mrs. Yerkes' findings, that somewhere between the seventh and the twelfth generations the animals entirely recovered from the effects of malnutrition and became normal again with respect to the relative weight of the central nervous system. They have remained normal in this regard up to the present time, as autopsies made at various periods on animals of the later generations have shown.

With the return of the central nervous system to its normal weight relations, the inbred rats must have regained much of their lost mental vigor, since in behavior tests animals of the fourteenth generation were found to be inferior to stock animals only in that they were slower and less active. The lesser activity of the inbred rats Mrs. Yerkes ascribes to "a greater timidity and a greater susceptibility to environmental conditions." Savageness, wildness, and timidity are heritable behavior complexes, according to R. Yerkes ('13), and since no attempt was made in the course of these experiments to eliminate these traits by selection, it is not surprising that they were manifested in a somewhat intensified form after many generations of close inbreeding.

3. DISCUSSION

Wherever inbreeding has been practiced it has usually been accused of producing anything and everything undesirable that has appeared in the offspring. The following quotation from Mitchell ('65) is quite typical of the belief that prevailed among zoologists, as well as among the laity, until the past decade, regarding the effects of consanguineous marriages:

Consanguinity in parentage tends to injure the offspring. This injury assumes various forms. It may show itself in diminished viability at birth; in feeble constitutions, exposing them to increased risks from the invasion of strumous disease in after life; in bodily defects and malformations; in deprivation or impairment of the senses, especially those of hearing and sight; and, more frequently than in any other way, in errors and disturbances of the nervous system, as in epilepsy, chorea, paralysis, imbecility, idiocy, and moral and intellectual insanity. Sterility or impaired reproductiveness is another result of consanguinity in marriage, but not one of such frequent occurrance as has been thought.

Stock breeders, also, have been imbued with the idea that inbreeding is always inimical to constitutional vigor and that it leads to sterility. For these reasons most of them have opposed the mating of animals related even in a remote degree. During the past few years it has been shown by a number of carefully controlled experiments that inbreeding does not necessarily produce the evil effects that have been attributed to it, and that the results obtained in any inbreeding experiment depend, primarily, on the soundness of the stock that is inbred; secondarily, on the selection of animals for breeding purposes, and, finally, on the environmental conditions under which the animals live. Haphazard inbreeding of inferior stock under unfavorable environmental conditions has produced many of the failures for which inbreeding alone has been held responsible.

Since the experiments of Crampe ('83), of Ritzema-Bos ('93, '94), and of von Guiata ('98, '00) have furnished the classic examples of the dire effects of inbreeding on rodents, it may be well to examine these experiments in some detail to see whether the unfavorable results obtained cannot be traced to some cause other than inbreeding per se.

Crampe's inbreeding experiments were begun, in 1873, with an Albino female and a white and gray male. From the mating of these rats he obtained the litter of five young which formed the basis of his breeding stock. These animals were inbred, in various degree of relationship, for seventeen successive generations. Crampe states that many of the animals were sterile and that others lost their reproductive instincts at the end of the first year. Various kinds of malformations appeared; the animals were seemingly too weak to resist disease of any kind, and they died at a relatively early age. The weakness of these rats and their susceptibility to disease, as well as the high degree of sterility among them, all point to the probability, as Ritzema-Bos suggests, that Crampe started his experiments with animals taken from a defective stock. Since results similar to Crampe's were obtained in the early part of my own experiments, I am inclined to the opinion that inadequate nourishment was a factor that was responsible, in great part, for his failure to maintain the stock in good physical condition.

Ritzema-Bos started his investigation in 1886 with a litter of twelve rats that was obtained from the mating of an Albino female and a wild Norway male. These rats were inbred, in various ways, for six years, during which time, Ritzema-Bos states, "about thirty generations were obtained." There is evidently some inaccuracy in this latter statement. The female albino rat does not cast her first litter until she is about three months old; wild rats do not breed, as a rule, before they are four or five months old. Assuming that all of the females used in Ritzema-Bos' experiments bred at the earliest possible age, i.e., three months, only four generations could possibly be produced in a year: this would give a maximum of twenty-four generations at the end of six years. In my own experiments an average of about three and one-half generations a year were obtained.

Ritzema-Bos gives data showing the average size of the litters and the number of infertile matings during the various years in which the work was in progress. These data have been reproduced in table 13.

During the first three years, as table 13 shows, there was little diminution in the average size of the litters produced. In the three following years, however, litter size decreased considerably, and at the end of the investigation the litters averaged less than one-half the size of those obtained in the beginning. These results certainly justify Ritzema-Bos' conclusion that: "Die fortgesetze Zucht in engster Verwandtschaft vermindert das Fortpflanzungsvermögen, kann sogar schliesslich vollkommene Unfruchtbarkeit verursachen." Lloyd ('12) has suggested that the deterioration in Ritzema-Bos' stock might have been due to overcrowding, since many varieties of rats will not breed in close confinement.

TABLE 13

Showing Ritzema-Bos' data for the average size of the litters and for infertile matings in a series of inbred rats

YEAR	AVERAGE NUMBER OF YOUNG PER LITTER	PER CENT INFERTILE MATINGS
1887	7.5	0.00
1888	7.1	2.63
1889	7.1	5.55
1890	6.5	17.39
1891	4.2	50.00
1892	3.2	41.18

Von Guaita obtained a number of white mice from a strain that had been inbred by August Weismann for twenty-nine generations. How these mice were inbred I do not know, since I have not been able to find any account of the details of this experiment. Von Guaita crossed these white mice with Japanese waltzing mice, and then inbred their descendants for five generations. The data regarding the average size of the litters obtained in these two sets of investigations are shown in table 14.

Weismann's data, given in table 14, show that the average size of the litters decreased directly as the inbreeding advanced, and so appear to indicate that inbreeding lessened the fertility of the mice. In this experiment there seems to have been a very great difference in the number of litters that were produced in the various generations. In the first two generations there was an average of about twenty-two litters to the generation: in the last nine generations the average was only about three litters to a generation. Such a small number of litters as that produced in the later generations of this series does not afford an opportunity for a careful selection of breeding stock, neither does it furnish sufficient data to make the results of statistical value.

In the successive generations of mice bred by von Guaita there was, to quote Davenport ('00): "a reduction in fertility of about 30 per cent, and this is probably due to close inbreeding." In order to make this deduction from von Guaita's data, however, it

TABLE 14

Showing the number and average size of the litters in twenty-nine generations of white mice inbred by August Weismann, and in seven generations of hybrid mice inbred by von Guaita

	GENERATIONS	NUMBER OF LITTERS	AVERAGE NUMBER OF YOUNG PER LITTER
Weismann's data for white mice.......	1–10	219	6.1
	11–20	62	5.6
	21–29	29	4.2
Von Guaita's data for hybrid mice.....	1	7	4.4
	2	15	3.0
	3	25	3.8
	4	31	4.3
	5	30	3.2
	6	11	2.3

is necessary to combine the records for three generations, as Davenport did. If the records for the various generations are considered separately, or grouped by twos, there is not the steady decrease in fertility with advancing inbreeding that Davenport's grouping of the data implies. Taking the data for the first two generations together, the average size of the litters was 3.7; for the next two generations there was an average of 4.0 young per litter; in the final group the litters averaged 2.7 young. Since the crossing of varieties is supposed to increase vigor and fecundity it seems strange that the F_1 and the F_2 litters in this series should contain a smaller average number of young than is found in the normal litter of either of the varieties that were crossed (five to six young per litter). Since crossing did not restore the normal fertility of the individuals, it would seem as if there must have been a strong tendency to sterility in each of the strains crossed. If such were the case, continued inbreeding, apparently without selection, would bring out this latent character and intensify it.

It seems rather remarkable that, of the many writers who have cited the results of the above series of experiments as proof that close inbreeding lessens fertility, not one, to my knowledge, has emphasized the fact that all of these experiments were made with hybrids and not with a pure strain. Hybridization in itself, as many investigators have noted, often produces a most marked effect on fertility. Some hybrids are equal, or even superior to the parent stock in fertility others are completely sterile; and among the hybrid offspring from various crosses all grades of productiveness from normal to complete sterility have been found. When hybridization increases fertility, its most marked effect is usually found in the animals of the F_1 and F_2 generations, and in later generations productiveness, as a rule, tends to decrease.

In connection with another problem, I have for several years been breeding the F_1 hybrids between the wild Norway and the albino rat, and I have also inbred various strains of 'extracted' rats, brother and sister, for several generations. Careful records have been kept of the litter production in all of these strains. While the great majority of the F_1 hybrid females are fertile, at least 25 per cent of the F_2 females are completely sterile, and about 10 per cent of those that do breed have only one or two litters. None of the 'extracted' strains that I have studied have even been as fertile as the inbred Albinos. The increase of sterility and the diminution in litter size with continued inbreeding has been very marked in some of these strains, but this lessened productiveness has been due, I believe, to hybridization, and it has not been influenced by inbreeding save in as far as inbreeding has intensified the tendencies which acted unfavorably upon productiveness. By rigid selection of only the most fertile individuals for breeding, from a large potential breeding stock, it might be possible to eliminate from the 'extracted' strains of rats the tendency to sterility that is seemingly caused by hybridization. Such a selection was not attempted, apparently, in any of the series of experiments cited above, nor was it done in my own work with hybrid stock. The experiments of Crampe, of Ritzema-Bos, and of von Guaita show unquestionably that fertility in hybrid rats is diminished by random inbreeding, but they cannot legitimately be used to give evidence regarding the effects of inbreeding on the fertility of a pure race.

Other series of inbreeding experiments made on pure strains of rodents show that inbreeding does not necessarily lead to a marked decrease in fertility. Neither Schultze ('03) nor Copeman and Parson ('09) found inbred mice less productive than the outbred strain; Castle ('16) did not find any great decrease in fertility in various races of rats inbred for seventeen generations. In the inbreeding experiment with guinea-pigs that has been carried on for several years at the Bureau of Animal Industry in Washington, there is, to quote Popenoe ('17): "no general deterioration. While a few strains have run out, others are nearly as vigorous as are the control families."

Results comparable to the above have been obtained with other animals. It is well known that inbreeding has been used extensively, and with very favorable results, in the building up of various strains of thoroughbred horses and cattle (Wriedt, '16), and the productiveness of these strains has not been greatly lessened. In the extensive series of inbreeding experiments with Drosophila, made by Castle et al. ('06), it was found that "inbreeding probably reduces very slightly the productiveness of Drosophila, but the productiveness may be fully maintained under constant inbreeding (brother and sister) if selection is made from the more productive families. . . . . Selection has a much greater influence on fertility than inbreeding, so that selection from the most productive pairs is able to more than offset the effects of inbreeding." The effectiveness of selection in increasing the fertility within an inbred strain is shown with great clearness in Moenkhaus' ('11) experiments with Drosophila. Moenkhaus was able to establish two distinct strains, one of high and one of low fecundity, by selecting, from among the variable offspring of the fourteenth generation of a closely inbred race, pairs of individuals showing very different degrees of productiveness and then inbreeding their descendants. Moenkhaus continued some of his lines for seventy-five generations and found that close inbreeding (brother and sister) was not deleterious either to fertility or to vigor. Hyde ('14) has found also that in certain strains of Drosophila sterility is an inherited character that is not influenced by inbreeding, and that "selection is an effective agent in controlling it."

In the present series of inbreeding experiments on the rat, the productiveness of the strain was decreased by malnutrition during the early generations, but normal fertility was restored as soon as the animals were adequately nourished. In later generations the fertility in the inbred animals was greater than that in the series of stock controls reared under similar environmental conditions. Thus even after a high degree of sterility had been introduced into the strain it was not retained in spite of the fact that close inbreeding was continued. In the later generations any tendency to sterility that appeared was evidently suppressed by selection. In the rat, as in Drosophila, selection seems a more potent factor for good than inbreeding is for evil.

During the past few years it has been shown, by a series of brilliant experiments, that characters tend to be inherited in groups and that this grouping depends upon the fact that the genetic factors involved are not segregated independently in gametogenesis, but tend to be linked together (Morgan et al., '15). In these experiments with the rat it has been found that animals that are large and vigorous when young tend to mature early, to be very productive, and to live to an advanced age. While all of these characters are influenced to a considerable degree by environmental conditions, it is evident that they must all depend to some extent upon heritable genes, since they are transmitted from generation to generation. A selection of breeding animals on the basis of size and early maturity has meant also selection for high fecundity and for characters that represent superior vigor of constitution, it would seem as if the genetic factors involved must tend to be inherited together, although they are probably not linked as are many of the genes in Drosophila.

Wentworth's ('13) experiments with Drosophila indicate that the supposed weaknesses from inbreeding are due to "the mere segregation of factors for lower vigor." Assuming that a similar segregation of these factors occurred in the inbred rats during the early generations, individuals containing the factors for 'lower vigor' were evidently eliminated by the selective action of malnutrition, and only those animals containing dominant genes for 'high vigor' were able to survive and to perpetuate their kind. Neither inbreeding nor selection is creative in its action. Selection can act on fertility only by preserving those individuals that contain genes for characters favorable to reproduction; inbreeding conserves these characters, and, to a certain extent, intensifies them. The action of both selection and inbreeding can be nullified by unfavorable conditions of environment or of nutrition which may produce a rapid deterioration in the fertility of any stock, regardless of the way in which the animals are bred.

It was shown by the work of Darwin ('78), as well as by a number of more recent experiments (Shull, '10; East and Hayes, '12; Hayes and Jones, '17), that crosses between different varieties of plants often produce hybrids that possess greater reproductive vigor than either parent stock. This result is due, according to East and Hayes ('12), to "the stimulation of vigor through heterozygosis." Inbreeding, these authors state, "tends to isolate homozygous strains which lack the physiological vigor due to heterozygosity. Decrease in vigor due to inbreeding lessens with decrease in heterozygosity and vanishes with the isolation of a completely homozygous strain." If the latter is a good strain, because of its gametic constitutional and natural inherent vigor, it is "ready to stand up forever under constant inbreeding."

The results obtained in these inbreeding experiments with the rat accord with the theory of East and Hayes to some extent. The effects of inbreeding on the fertility and on the vigor of the rats were obscured in the early generations by the action of malnutrition, but it would appear that the animals lost very little of their constitutional vigor during this time, since adequate nutrition soon restored the normal productiveness of the strain and its general vigor as well. Apparently at about the tenth generation, the inbred rats became sufficiently homozygous for vigor to become fairly constant. Beyond the point, as the data show, there was little variation in the fertility or in the longevity of the animals up to the twenty-fifth generation. Selection and favorable environment kept the strain at a point of high productiveness, but, under the conditions of the experiment, they did not increase vigor beyond the stage which was reached at the tenth generation. As already stated, no attempt was made in the course of these experiments to influence fertility by selecting breeding animals from large or from small litters. Whether selection can act in the rat, as it does in Drosophila, and produce strains of high and of low productiveness within a line that has been inbred for many generations is a problem for the future. As the strain has been very fertile for many generations it seems very improbable that any sudden loss in fertility will occur in the future, unless sterility appears as a mutation which cannot be eliminated by selection.

While corresponding records for the two inbred series (A,B) are in close agreement, there are, nevertheless, differences between the series that have persisted from the very beginning. Female A, one of the two females with which the experiment were started, showed a relatively high degree of fertility since she gave birth to five litters, containing thirty-five young, before she was killed at the age of one year: female B, a sister of female A, cast only one litter of five young, although she was paired continuously for several months and appeared to be in good physical condition. The two litter brothers with which these females were paired showed no marked differences in size or in vigor. The rats of the A series (which were descended from female A) were, as the records show, somewhat more fertile than the rats of the B series (the descendants of female B), and they also tended to mature earlier and to live longer. The differences found were not very marked in any case, and they might well be ignored were it not for the fact that in all of the characters noted the animals of the A series were superior to those of B series. Environment cannot be held accountable for these differences, since the two series of inbreds were kept constantly under similar conditions of light, of temperature, and of nutrition. Although the two series were descended from the same ancestral stock, apparently there was an inherent difference in the gametic constitution of the two pairs of rats with which the experiment was started, which persisted from generation to generation and produced the effects noted.

While the inbred strain of rats that has been developed in the course of these experiments is seemingly superior to the average run of stock Albinos in body size, in fertility, and in longevity, I do not claim that this superiority is due solely to the fact that the animals were inbred, neither do I wish to assert that, in general, inbreeding is better than outbreeding for building up and for maintaining the general vigor of a race. The two forms of breeding are not mutually exclusive: each has its merits, and the one should supplement the other to bring out the best in any stock. The favorable results that have been obtained in these experiments have been achieved through the constant selection of only the best animals from a larger number available for breeding purposes and by keeping the environmental conditions as uniform and as favorable as it was possible to make them. These experiments have fully demonstrated, I think, that even in mammals the closest form of inbreeding possible, i.e., the mating of brother and sister from the same litter, is not necessarily injurious either to the fertility or to the constitutional vigor of a race even when continued for many generations. Success or failure in inbreeding experiments depends chiefly, it would seem, on the character of the stock that is inbred, on the manner in which the breeding animals are selected, and on the environmental conditions under which the animals are reared. There is no warrant, therefore, either in theory or in fact, for the dogmatic assertion of Kraemer ('13) that: "continued inbreeding must always result in weakened constitution, through its own influence."

4. SUMMARY

1. The present paper gives data showing the fertility, the time of puberty, and the longevity in two series of albino rats (A,B) that were inbred, litter brother and sister, for twenty-five generations.

2. Data given for the A series of inbreds comprise 1752 litters containing 13,116 individuals, or an average of 7.5 young per litter (table 1); records for the B series of inbreds include 1656 litters having a total of 12,336 members, or an average of 7.4 young per litter (table 2). The two series combined comprise a total of 3408 litters which contained 25,452 individuals. For the entire strain the average size of the litter was 7.5 young.

3. In any litter series of albino rats, whether the animals are inbred or outbred, the first litter cast is the smallest of the series, as a rule; the second litter is the largest; while the third and fourth litters are about the same size and a little smaller than the second litter.

4. The size of any litter cast depends chiefly on the age and physical condition of the female, and is not affected by the relatedness or the unrelatedness of the parents.

5. A comparison of the data for the inbred strain with data for litter size obtained from a series of stock Albinos reared under the same environmental conditions as the inbred strain shows that each litter of the stock series was relatively smaller than the corresponding litter in the inbred group. For the entire series of 424 stock litters the average size was 6.7 young per litter. This average is 0.8 less than the average for litter size in the inbred strain (table 7).

6. In the A series of inbreds the range in litter size was from one to seventeen; in the B series it was from two to fifteen. In both series the most frequent litter size was seven (table 8).

7. In the early generations of these inbred rats malnutrition greatly delayed the time of puberty in the animals. In the later generations, under favorable nutritive conditions, the animals bred at a relatively early age.

8. While the records give no definite information regarding the number of sterile animals in the inbred strain, they show clearly that inbreeding did not decrease the productiveness of the animals. Of the 954 females that were used for breeding, 653, or 68.5 per cent cast the required number of four litters. Where partial sterility occurred in apparently healthy females it was found to be due to a diseased condition of the reproductive organs.

9. The constitutional vigor of these rats was apparently not impaired to any extent by inbreeding. Only two kinds of malformations were found in the animals of the inbred strain after food conditions were improved: one individual was born tailless and about a dozen individuals lacked one or both eyeballs. Both of these defects occur in outbred stock Albinos and neither appears to be heritable.

10. Under the conditions of these experiments the span of life in both the males and females in each of the inbred series was increased. The records show that inbred males tended to live longer than did inbred females: a reversed relation was found in the animals of the control series. In the inbred colony as a whole, the females seemed to be longer lived than the males and they were less susceptible to disease at all ages.

11. According to the behavior tests that were made, inbred Albinos are slower, less active, more timid and nervous, and somewhat more savage than stock Albinos that are outbred.

12. High fecundity, early sexual maturity, and vigorous growth are characters that seemed to be inherited as a group in the inbred strain of rats. It seems probable that the genetic factors on which these characters depend do not segregate independently, but tend to combine in gametogenesis.

13. The animals of the A series were slightly more fertile than the animals of the B series, they attained sexual maturity earlier, as a rule, and they lived longer. These differences probably depended in some way on a dissimilarity in the gametic constitution of the two pairs of individuals with which the experiments were started.

14. The results obtained in these experiments do not accord with the general view regarding the effects of inbreeding, since they indicate that inbreeding per se is not necessarily inimical either to fertility or to vigor. Success or failure in any series of inbreeding experiments would seem to depend on the character of the stock that is inbred, on the manner in which breeding animals are selected, and on the environmental conditions under which the animals are reared.

STUDIES ON INBREEDING

III. THE EFFECTS OF INBREEDING, WITH SELECTION, ON THE SEX RATIO OF THE ALBINO RAT

HELEN DEAN KING

The Wistar Institute of Anatomy and Biology

ONE FIGURE

During the latter part of the nineteenth century it was generally believed that sex in man and in various animals is determined mainly by the amount of nourishment that the embryos receive; well nourished embryos were supposed to become females; those that were poorly nourished were assumed to develop into males. A considerable amount of evidence in favor of this view was collected by Düsing ('83, '84, '86), who maintained, furthermore, that close inbreeding interferes with embryonic nutrition, by lessening the vitality of the mother, and so produces a great excess of male young.

In the literature of the succeeding twenty years that deals with the subject of sex determination, Düsing's statement regarding the effect of inbreeding on the sex ratio was widely quoted and generally credited. Those who challenged the truth of the assertion were, in the main, advocates of the ancient theory, generally ascribed to Hippocrates (460–377 B.C.), that sex is determined in the ovary; eggs from the right ovary producing males and those from the left ovary developing into females. During this period three series of experiments were made that give data regarding the sex-proportions in a closely inbred stock. Huth ('87) inbred rabbits, brother and sister, for six generations and found a relatively low sex ratio (78.8 ♂ : 100 ♀) among the ninety young in which the sex was ascertained; Copeman and Parsons ('04) obtained a similar result in their inbreeding experiments with mice. Schultze ('03) concluded that inbreeding

99

has no pronounced tendency to produce an excess of male young, although he found a high sex ratio (110.9♂: 100 ♀) among 135 mice that were the offspring of brother and sister matings. .The question as to whether inbreeding does or does not alter the sex ratio was not satisfactorily answered by any of these experiments, for in each case the number of animals used was small, and there was, apparently, no selection of the best stock for breeding or any way of checking the results. Moreover, none of these investigations were continued long enough to give evidence that could be considered as conclusive.

The effects of inbreeding on the sex ratio seemed to me to be a problem of sufficient importance to warrant a careful and prolonged investigation. For if it were possible to swing the sex ratio of any animal in a definite direction by factors that could be controlled, one might hope to gain valuable information regarding the nature of sex—a problem that has been a favorite subject of speculation for many centuries and one that modern methods of research have not, as yet, satisfactorily solved.

1. MATERIAL, METHOD, AND SCOPE OF THE INVESTIGATION

The albino rat (Mus norvegicus albinus) was the animal used in this investigation, which was begun in 1909. Details regarding the manner in which the experiments were conducted were given in the first paper of this series (King, '18), but it has seemed advisable to repeat them here in order to give a clear understanding of the way in which the problem has been approached.

The basis of the inbred strain was a litter of four albino rats, two males and two females, taken from the general colony of these animals maintained at The Wistar Institute of Anatomy and Biology in Philadelphia. The litter was selected for the purpose in view solely because of its size, not because of the ancestry or the vigor of the animals. One of the two females in the litter was called 'A', and her descendants form the A series of inbreds; the other female was called 'B', and her descendants are the B series of inbreds.

Since the mating of brother and sister from the same litter is the closest form of inbreeding possible in mammals, such matings would be expected to be more potent than any other kind in producing an alteration in the sex ratio. In these experiments, therefore, brother and sister matings only were used to obtain strictly inbred litters from which all females used for breeding were taken. The plan of breeding that was followed through the first twenty-five generations of these animals was this: Females A and B, as well as all of the females in their respective lines that were subsequently used for breeding, were paired twice with a litter brother and then twice with an unrelated male taken from the stock colony. Sex records for the first two litters produced by any group of females might be expected to show whether inbreeding had any effect on the sex ratio; sex data for the third and for the fourth litters cast by these same females would, it was hoped, indicate whether the male or the female was responsible for the alteration, if any, in the sex ratio. For convenience the litters obtained from the mating of inbred females with stock males are here designated as 'half-inbred' litters; no animals from such litters have ever been reared.

Emphasis should be placed on the fact that, with few exceptions, the sex data given in this paper were obtained by examining the litters very soon after their birth. The sexes can readily be distinguished at this time, as Jackson ('12) has shown, and if accurate sex data are wanted it is imperative that they be taken as soon as possible, since the young that are stillborn, or those that die soon after birth, are usually eaten by the mother within a few hours.

In order to keep track of a large series of animals it was necessary to find some way in which the pedigree of any particular individual could be told by a glance at the record card. The scheme of marking devised, which is outlined below, has proved to be very convenient and also most satisfactory for the filing of permanent records. The letter A or B is used to show from which of the two females, A or B, the animal was descended, and thus places the individual in its proper series. The serial letter is preceded in all cases by a number which signifies the generation to which the animal belonged. An index number, 2, 3, or 4, following the serial letter shows in which of the mother's litters the animal was born; if no index number is used the rat was a member of its mother's first litter. The subscript following the serial letter is the number which serves to distinguish each particular rat from the other rats belonging to the same generation and litter group. When it is desired to indicate the sex of the individual its number is inclosed by the sex symbol. An illustration will, perhaps, render the scheme clearer.

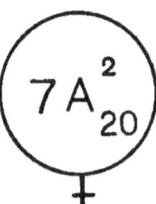

This symbol denotes a female rat belonging in the seventh generation of the A series of inbreds. She was a member of the second litter cast by her mother, and her individual number in the series of rats belonging to the second litters of the seventh generation was twenty.

In the early generations of both inbred series the animals suffered severely from malnutrition which produced a marked effect on their growth, fertility, and longevity, as previous papers in this series have shown (King, '18, '18 a). During this period a considerable proportion of the individuals were sterile, and it was not possible to select animals for breeding; any rats that would breed at all were used to continue the strain. Nutritive conditions were improved at the time that the rats of the fourth inbred generation were approaching maturity, and a decided improvement in the condition of the animals was noted in a very short time: they gained rapidly in weight, the litters cast became larger and sterility almost disappeared. At this stage of the investigation it became possible to attempt to alter the sex ratio by selection within the inbred strain. From the seventh generation on, every female in the A series of inbreds that was used for breeding was taken from a litter that contained an excess of males; breeding females in the B series of inbreds were all taken from litters containing an excess of females. The plan of pairing a female twice with a litter brother and then twice with an unrelated stock male was continued through the first twenty-five generations of both inbred series.

In each series litters having the desired sex ratio were reared as possible breeding stock only when the young were of large size and lusty at birth; all other litters were discarded regardless of their sex ratio. At the time that the animals became sexually mature the largest and most vigorous pairs were the ones taken to continue the strain. Selection of breeding stock, it will be noted, was based primarily on the sex ratio in the litters, not on the size or on the vigor of the young. This means that the animals in one generation that became the progenitors of the succeeding generation were selected because of their parents, tendency to produce young of a certain sex. A pair of rats that produced two litters, each of which had the desired sex ratio, was considered as having an unusually strong tendency to produce unisexual young; individuals from each of these litters were used for breeding when possible. The basis of the selection, therefore, was along the line in which Pearl ('12, '12 a, '17) has obtained such marked success in increasing egg production in poultry, i.e., according to the ability of the parents to transmit to the offspring the quality desired.

In the early part of this investigation the number of breeding females was, of necessity, small, but in the later generations about twenty females in each series were used for breeding, so that at least 1000 rats were obtained in each generation of the inbred strain. Sex records for the first twenty-five generations are given in the present paper; the data comprise 3408 litters containing 25,452 individuals.

2. THE NORMAL SEX RATIO IN THE ALBINO RAT

The normal sex ratio in any species can properly be determined only by obtaining the sex data for the total number of offspring produced by many females during the entire period of their reproductive activity. Unfortunately, no such series of data for the albino rat have been recorded, and only two sets of observations regarding the normal sex ratio in this animal have, as yet, appeared. Cuénot ('99) examined thirty litters of albino rats, containing 255 young, and found a sex ratio of 105.6♂ : 100♀ ; data for 1089 litters of stock Albinos, collected by King and Stotsenburg ('15), gave a sex ratio of 107.5♂ : 100♀. Neither of these determinations seemed to furnish a proper standard for comparison with the sex ratios obtained in the inbred strain, even though they differed by less than two points. The number of individuals examined by Cuénot was too small to give results of much statistical value. The sex ratio given by King and Stotsenburg was based on the findings for a relatively large number of animals, but the litters recorded were, for the most part, cast by females that had not reached the height of their reproductive activity. The sex ratio among the offspring of young females could not justly be taken as a norm for the Albino strain in general, since it has been shown that in the albino rat the sex of the young seemingly depends, to a certain extent, on the age of the mother (King, '16 a).

In order to ascertain the normal proportion of the sexes in the strain of Albinos from which the inbred animals were taken, I obtained the complete breeding history of a considerable number of stock females during the past four years. As all of these individuals were reared under the same environmental conditions as the inbred rats, the sex ratio among their young would seem to be a suitable standard by which to judge the sex ratios found in various generations of the inbred animals. To make the ratios more strictly comparable, the data for only the first four litters of the stock series were used in computing the sex ratio which was to serve as the norm. These data, arranged by litter groups, are shown in table 1.

TABLE 1

Showing the sex ratios in the first four litters of a series of stock albino rats

LITTER SERIES	NUMBER LITTERS	NUMBER INDIVIDUALS	MALES	FEMALES	NUMBER MALES TO 100 FEMALES
1	116	717	385	332	115.9
2	116	843	426	417	102.2
3	103	671	328	343	95.6
4	89	587	302	285	105.9
	424	2818	1441	1377	104.6

Table 1 shows that there was a relatively large excess of males in the first litters cast by this series of stock females (115.9♂ : 100 ♀), and that in succeeding litters the sex ratio tended to fall considerably. A similar change in the sex ratio of successive litters of mice was noted by Copeman and Parsons ('04), and was found also by King and Stotsenburg ('15; table 7) in a series of stock albino rats. Large groups of statistics for human births, as summarized by Ahlfeld ('76), by Düsing ('83, '84), by Punnet ('03), and by Newcomb ('04), all show that the sex ratio is very high among the first children of young mothers and then tends to fall with succeeding births until the mother is about thirty years old. Whether a similar change in the sex ratio is characteristic of other mammals has not been determined as yet.

Among the 2818 individuals comprised in this series of stock litters there were 104.6 males to each 100 females. A sex ratio of 105♂ : 100 ♀ was, therefore, taken as the norm by which to judge the sex ratios obtained in the various groups of inbred rats. This sex ratio, it will be noted, is very close to that given by Cuénot, and is lower, by over two points, than the sex ratio found in the large group of stock Albinos born in The Wistar Institute colony during the years 1911–1914 (King and Stotsenburg, '15).

3. THE SEX RATIO IN INBRED LITTERS OF ALBINO RATS

The A series of inbred rats may be designated as the 'male line,' since after the sixth generation all of the breeding females in this series were taken from litters that contained an excess of males. Table 2 gives, by litter groups, the sex data for the 13,116 individuals obtained in the first twenty-five generations of this series.

Table 2 is inserted chiefly for reference, and a detailed analysis of the data, as given, will not be attempted. The summary of the data for the various litter groups shows that the sex ratio for the first litters produced was much higher than that for the second, third, and fourth litters. A similar change in the sex ratio was noted in the litter series of stock animals given in table 1.

The B series of inbreds is called the 'female line,' since, after the sixth generation, all breeding females in this series came from

TABLE 2

Showing the number of individuals and the sex ratios in each of the first twenty-five generations of the A series of inbred
(data arranged in litter groups)

GENERATION	FIRST LITTER (INBRED)					SECOND LITTER (INBRED)					THIRD LITTER (HALF INBRED)					FOURTH LITTER (HALF-INBRED)				
	Number of litters	Number of individuals	Males	Females	Number of males to 100 females	Number of litters	Number of individuals	Males	Females	Number of males to 100 females	Number of litters	Number of individuals	Males	Females	Number of males to 100 females	Number of litters	Number of individuals	Males	Females	Number of males to 100 females
1	1	7	3	4	75.0	1	7	3	4	75.0	1	6	6			1	9	7	2	350.0
2	3	20	11	9	122.2	4	27	11	16	68.8	3	19	12	7	171.4	3	14	7	7	100.0
3	7	35	23	12	52.1	5	26	15	11	136.4	5	27	15	12	125.0	5	21	9	12	75.0
4	13	58	36	22	163.6	12	77	37	40	92.5	8	52	29	23	126.1	6	46	28	18	155.6
5	18	117	63	54	116.7	18	130	60	70	85.8	15	111	52	59	88.1	10	58	25	33	75.8
6	15	94	44	50	88.0	15	102	55	47	117.0	14	86	49	37	132.4	11	63	32	31	103.2
7	16	101	64	37	173.0	16	129	74	55	134.5	15	109	64	45	142.2	9	61	33	28	117.9
8	17	122	61	54	110.3	17	142	79	63	125.4	15	100	57	43	132.6	8	56	29	27	107.4
9	17	105	51	54	94.4	17	123	71	52	136.5	16	121	61	60	101.7	12	76	37	39	94.9
10	20	131	71	60	118.3	20	156	76	80	95.0	20	161	76	85	89.4	17	137	76	61	124.6
11	21	144	82	62	132.3	21	170	104	66	157.6	20	164	87	77	113.0	18	135	68	67	101.5
12	20	145	90	55	163.6	20	154	85	69	123.2	19	152	75	77	97.4	17	140	78	62	125.8
13	22	140	86	54	116.3	22	180	93	87	106.9	21	156	91	65	140.0	20	157	92	65	141.5
14	21	139	81	58	139.7	21	188	95	93	97.9	21	182	105	77	136.4	18	136	73	63	115.9
15	23	171	109	62	175.8	23	195	103	90	114.4	21	173	90	83	108.4	17	136	68	68	100.0
16	21	143	78	65	120.0	21	165	94	71	132.4	18	121	69	52	132.7	10	62	36	26	138.5
17	27	210	116	94	124.0	27	244	125	119	105.0	25	213	109	104	104.8	23	178	101	77	131.2
18	23	168	97	71	136.6	23	197	104	93	111.8	19	164	85	79	107.6	15	110	63	47	134.0
19	25	147	89	58	153.4	23	178	97	81	119.8	22	172	90	82	109.8	15	118	68	50	136.0
20	27	198	113	85	141.3	27	224	118	106	111.3	23	149	84	65	129.2	17	130	73	57	128.1
21	27	199	107	92	116.3	27	217	119	98	121.5	26	223	114	109	104.6	22	187	95	92	103.3
22	27	208	118	90	135.8	27	234	123	111	110.9	24	218	116	102	113.7	17	113	52	61	85.2
23	25	186	112	74	151.4	25	183	103	80	128.8	22	161	87	74	117.6	19	142	80	62	135.5
24	25	184	100	84	119.0	25	200	108	92	117.4	22	177	84	93	90.3	21	140	77	63	122.2
25	26	173	93	80	116.3	26	203	108	95	113.8	23	166	95	71	133.8	15	104	59	45	121.1
	185	3353	1890	1465	129.0	483	3849	2058	1791	114.9	438	3383	1802	1581	113.9	346	2529	1366	1163	117.4

litters containing an excess of females.   Reference data showing the proportion of males and females produced in the various generations of this series are given, by litter groups, in table 3. The data comprise a total of 1656 litters containing 12,336 individuals.

The summary for each of the four litter groups of the B series (table 3) shows that the sex ratio was at its lowest point in the first litter group, and then tended to rise in each of the subsequent groups.   This is a reversed relation of the sex ratios to that shown in the litters of the stock controls (table 1) and in the litter groups of the A series (table 2), and would seem to indicate that some agency, other than environment or the age of the mother, had influenced the relative proportion of the sexes in this series of animals.

In order to compare the sex ratios in the litters sired by inbred males with the sex ratios in the litters sired by stock males, the sex records for the first and second litters produced in each generation of the two series were combined, as were also the records for the third and fourth litters.   Table 4 shows the combined data for the litter groups of the A series; table 5 shows similar data for the litter groups of the B series.

Reference to the data given in table 4 and in table 5 will be made later.

To facilitate an analysis of the results obtained in the A series of inbreds, the data, as shown in table 4, were combined in generation groups (table 6).   This grouping of the data was purely arbitrary.   It seemed useless to compare such large series of records generation by generation, or even to combine the records for two succeeding generations.   Since after the sixth generation the selection of breeding animals was made according to a definite plan, it would seem that, logically, the data for the first seven generations should form one group.   Such a group, however, was too large for the purpose of ascertaining whether selection produced a varying effect in different generations.   It was finally decided to make a total of eight groups, each of which, except the first, should contain the data for three generations. Because of the small number of individuals, records for the first four generations were combined in one group.

TABLE 8

Showing the number of individuals and the sex ratios in each of the first twenty-five generations of the B series of inbred

(data arranged in litter groups)

GENERATIONS	FIRST LITTER (INBRED)					SECOND LITTER (INBRED)					THIRD LITTER (HALF-INBRED)					FOURTH LITTER (HALF-INBRED)				
	Number of litters	Number of individuals	Males	Females	Number of males to 100 females	Number of litters	Number of individuals	Males	Females	Number of males to 100 females	Number of litters	Number of individuals	Males	Females	Number of males to 100 females	Number of litters	Number of individuals	Males	Females	Number of males to 100 females
1	1	5	2	3	66.7															
2	2	12	6	6	100.0	2	19	6	13	68.4	2	17	9	8	112.5					
3	7	36	19	17	111.8	5	26	15	11	136.4	2	9	5	4	125.0	1	9	3	6	50.0
4	11	66	33	33	100.0	8	69	41	28	146.4	7	56	27	29	93.1	3	18	9	9	100.0
5	20	133	66	67	98.6	20	158	82	76	107.9	18	150	71	79	89.9	13	95	51	44	115.9
6	15	106	59	47	125.5	15	118	61	57	107.0	14	102	51	51	100.0	8	74	41	33	124.2
7	15	79	42	37	113.5	15	105	54	51	105.9	15	104	49	55	89.1	10	78	37	41	90.2
8	15	108	46	62	74.2	15	129	69	60	115.0	13	94	48	46	104.4	3	21	10	11	90.9
9	20	126	62	64	96.8	20	183	90	93	96.8	17	127	68	59	115.3	10	49	26	23	113.0
10	17	104	48	56	85.7	17	145	67	78	85.9	16	107	53	54	98.1	14	93	47	46	102.2
11	19	112	49	63	77.8	19	128	66	62	106.5	18	138	60	78	76.9	18	149	78	71	109.9
12	20	139	66	73	90.4	20	175	92	83	110.8	19	154	71	83	85.5	12	68	29	39	74.1
13	21	154	54	100	54.0	21	168	73	95	76.8	19	150	70	80	87.5	13	104	50	54	92.6
14	21	142	57	85	67.1	21	135	68	67	101.5	20	161	84	77	109.1	18	132	58	74	78.4
15	20	131	61	70	87.1	20	138	58	80	72.5	20	173	88	85	103.5	17	140	72	68	105.9
16	21	158	57	101	56.4	24	184	85	99	85.9	22	190	76	114	66.7	19	132	72	60	120.0
17	22	158	68	90	75.6	22	170	74	96	77.1	21	173	83	90	92.2	17	113	52	61	85.2
18	23	168	76	92	82.6	23	180	83	97	85.6	20	171	82	89	92.1	20	149	70	79	88.6
19	24	186	73	113	64.6	24	221	98	123	79.7	22	195	93	115	91.2	13	96	34	62	54.8
20	26	174	76	98	77.6	26	179	83	96	86.5	21	157	80	77	103.9	11	88	44	44	100.0
21	24	176	77	99	77.8	24	189	91	98	92.9	18	138	64	74	86.5	16	112	50	62	80.6
22	26	174	87	87	100.0	26	210	91	119	76.5	24	211	96	115	92.2	19	139	67	92	93.1
23	22	181	82	99	82.8	22	195	83	112	74.3	20	145	65	82	77.1	15	126	53	73	72.6
24	27	207	96	111	86.5	27	189	76	113	67.3	26	175	90	85	105.9	17	114	58	56	103.6
25	26	184	83	101	82.2	26	193	86	107	80.4	25	179	88	91	96.7	20	136	62	74	83.8
	468	3219	1445	1774	81.5	462	3606	1692	1914	88.3	413	3276	1569	1707	91.3	307	2235	1073	1162	92.3

TABLE 4

Showing the sex ratios in the inbred and in the half-inbred litters produced in each of the first twenty-five generations of the A series of inbreds

GENERATIONS	INBRED (FIRST AND SECOND LITTERS)					HALF-INBRED (THIRD AND FOURTH LITTERS)					SUMMARY OF ALL LITTERS				
	Number of litters	Number of individuals	Males	Females	Number of males to 100 females	Number of litters	Number of individuals	Males	Females	Number of males to 100 females	Number of litters	Number of individuals	Males	Females	Number of males to 100 females
1	2	14	6	8	75.0	2	15	13	2	650.0	4	29	19	10	190.0
2	7	47	22	25	88.0	6	33	19	14	135.0	13	80	41	39	105.1
3	12	61	27	34	79.1	10	48	24	24	100.0	22	109	51	58	87.9
4	25	135	73	62	117.4	14	98	57	41	139.0	39	233	130	103	126.2
5	36	247	123	124	99.2	25	169	77	92	83.7	61	416	200	216	92.6
6	30	196	99	97	102.1	25	149	81	68	119.2	55	345	180	165	110.0
7	32	230	138	92	150.0	24	170	97	73	132.9	56	400	235	165	142.4
8	34	264	143	121	118.2	23	156	86	70	122.9	57	420	229	191	119.9
9	34	228	122	106	115.0	28	197	98	99	99.0	62	425	220	205	107.3
10	40	287	147	140	105.0	37	298	152	146	104.1	77	585	299	286	104.5
11	42	314	186	128	145.3	38	299	155	144	107.6	80	613	341	272	121.9
12	40	299	175	124	141.1	36	292	153	139	110.2	76	591	328	263	124.7
13	44	340	179	161	111.2	41	313	183	130	140.8	85	653	362	291	124.4
14	42	327	174	153	113.7	39	318	178	140	127.1	81	645	352	293	120.1
15	46	364	212	152	139.5	38	309	158	151	104.6	84	673	370	303	122.1
16	42	308	172	136	126.5	28	183	105	78	134.6	70	491	277	214	129.7
17	54	454	241	213	113.1	48	391	210	181	116.0	102	845	451	394	114.5
18	46	365	201	164	122.6	34	274	148	126	117.5	80	639	349	290	120.0
19	46	325	186	139	133.8	37	290	158	132	119.7	83	615	344	271	126.9
20	54	417	231	186	123.1	40	279	157	122	128.7	94	696	388	308	126.0
21	54	416	226	190	118.9	48	410	209	201	104.0	102	826	435	391	111.3
22	54	437	241	196	123.0	41	331	168	163	103.1	95	768	409	359	113.9
23	50	369	215	154	139.6	41	303	167	136	122.8	91	672	382	290	131.7
24	50	384	208	176	118.2	43	317	161	156	103.2	93	701	369	332	111.1
25	52	376	201	175	114.9	38	270	154	116	132.8	90	646	355	291	122.0
	968	7204	3948	3256	121.3	784	5912	3168	2744	115.5	1752	13116	7116	6000	117.4

As the number of individuals in each of the first seven generations of the A series was comparatively small, it is not surprising that the sex ratios in the inbred and in the half-bred groups of litters should show a wide range of variation (table 4). When the records for these generations were combined, as shown in table 6, it was found that the 144 inbred litters had a sex ratio of 110.4♂ :

TABLE 5

Showing the sex ratios in the inbred and in the half-inbred litters produced in each of the first twenty-five generations of the B series of inbred

GENERATIONS	INBRED (FIRST AND SECOND LITTERS)					HALF-INBRED (SECOND AND THIRD LITTERS)					SUMMARY OF ALL LITTERS				
	Number of litters	Number of individuals	Males	Females	Number of males to 100 females	Number of litters	Number of individuals	Males	Females	Number of males to 100 females	Number of litters	Number of individuals	Males	Females	Number of males to 100 females
1	1	5	2	3	66.7						1	5	2	3	66.7
2	4	31	12	19	63.2	3	26	12	14	85.7	7	57	24	33	72.8
3	12	62	34	28	121.4	2	9	5	4	125.0	14	71	39	32	121.9
4	19	135	74	61	121.3	10	74	36	38	94.7	29	209	110	99	111.1
5	40	291	148	143	103.5	31	245	122	123	99.2	71	536	270	266	101.5
6	30	224	120	104	115.4	22	176	92	84	109.5	52	400	212	188	112.9
7	30	184	96	88	109.1	25	182	86	96	89.6	55	366	182	184	98.9
8	30	237	115	122	94.3	16	115	58	57	101.8	46	352	173	179	96.6
9	40	309	152	157	96.8	27	176	94	82	114.6	67	485	246	239	102.9
10	34	249	115	134	85.8	30	200	100	100	100.0	64	449	215	234	91.9
11	38	240	115	125	92.0	36	287	138	149	92.6	74	527	253	274	92.3
12	40	314	158	156	101.3	31	222	100	122	82.0	71	536	258	278	92.8
13	42	322	127	195	65.1	32	254	120	134	89.6	74	576	247	329	75.1
14	42	277	125	152	82.2	38	293	142	151	94.0	80	570	267	303	88.1
15	40	269	119	150	79.3	37	313	160	153	104.5	77	582	279	303	92.1
16	48	342	142	200	71.0	41	322	148	174	85.1	89	664	290	374	77.5
17	44	328	142	186	76.3	38	286	135	151	89.4	82	614	277	337	82.2
18	46	348	159	189	84.1	40	320	152	168	90.5	86	668	311	357	87.1
19	48	407	171	236	72.5	35	291	127	164	77.4	83	698	298	400	74.5
20	52	353	159	194	82.0	32	245	124	121	102.5	84	598	283	315	89.8
21	48	365	168	197	85.3	34	250	114	136	83.8	82	615	282	333	84.7
22	52	384	178	206	86.4	43	350	163	187	87.2	95	734	341	393	86.8
23	44	376	165	211	78.2	45	271	116	155	74.8	79	647	281	366	76.8
24	54	396	172	224	76.8	43	289	148	141	105.0	97	685	320	365	87.7
25	52	377	169	208	81.3	45	315	150	165	90.9	97	692	319	373	85.5
	930	6825	3137	3688	85.1	726	5511	2642	2869	92.1	1656	12336	5779	6557	88.1

100 ♀, and that the 106 half-inbred litters had an even higher proportion of males (114.0 ♂ : 100 ♀). For the total of 250 litters the sex ratio was 113.2 ♂ : 100 ♀.

Until the seventh generation, as already stated, there was no selection of breeding animals in either series. As the sex ratio among the animals in the early generations of the A series was

TABLE 6

Showing, by generation groups, the sex ratios in the inbred and in the half-inbred letters of the A series (male line)

GENERATION GROUPS	INBRED (FIRST AND SECOND LITTERS)					HALF-INBRED (THIRD AND FOURTH LITTERS)					SUMMARY OF ALL LITTERS				
	Number of litters	Number of individuals	Males	Females	Number of males to 100 females	Number of litters	Number of individuals	Males	Females	Number of males to 100 females	Number of litters	Number of individuals	Males	Females	Number of males to 100 females
1–4	46	257	128	129	99.2	32	194	113	81	139.5	78	451	241	210	114.8
5–7	98	673	360	313	115.0	74	488	255	233	109.0	172	1161	615	546	112.6
1–7	144	930	488	442	110.4	106	682	368	314	114.0	250	1612	856	756	113.2
8–10	108	779	412	367	112.5	88	651	336	315	106.6	196	1430	748	682	109.7
11–13	126	953	540	413	130.7	115	904	491	413	118.8	241	1857	1031	826	124.8
14–16	130	999	558	441	126.5	105	810	441	369	121.9	235	1809	999	810	123.3
17–19	146	1144	628	516	121.7	119	955	516	439	117.5	265	2099	1144	955	119.7
20–22	162	1270	698	572	122.0	129	1020	534	486	109.8	291	2290	1232	1058	116.4
23–25	152	1129	624	505	123.5	122	890	482	408	118.1	274	2019	1106	913	121.1
8–25	824	6274	3460	2814	122.3 ±1.55	678	5230	2800	2430	115.6 ±1.47	1502	11504	6260	5244	119.3 ±1.36

some eight points above the norm, it might appear that inbreeding had tended to increase the relative number of males. Such an interpretation of the results is not warranted, however, since the sex ratio in the litters produced by the mating of unrelated parents was higher than that in the litters obtained by the mating of brother and sister, and since a similar increase in the sex ratio was not found in corresponding litters of the B series (table 7).

As the females of the seventh generations that were used for breeding were all taken from litters that contained an excess of males, it is among their offspring that we may look for a possible alteration of the sex ratio as a result of selection. The sex ratio in the inbred litters of the eighth generation of the A series was 118.2♂ : 100 ♀. This sex ratio is very much lower than that found in the inbred litters of the seventh generation (150♂ : 100 ♀), but it is still 13 points above the norm (105♂ : 100 ♀). As examination of the records given in table 4 shows that *in only one generation (the tenth) after the eighth did the sex ratio for the inbred litters fall to norm, in all other generations it was considerably above the norm,* the highest ratio (145.3 ♂ : 100 ♀) being found in the litters of the eleventh generation.

While the sex ratios for the inbred litters of the eighth to the twenty-fifth generations varied considerably, the variation was much less after the twelfth generation than before (table 4). A part of this variation was doubtless phenotypic, since seasonal changes in temperature seem to alter the sex ratio in the rat (King and Stotsenburg, '15), and probably also other agencies, such as the age of the mother (King, '16 a), have a similar effect. As all of the sex ratios were relatively high, however, the deviations from the norm cannot be ascribed either to environment or to chance, so they must have been due, in part at least, to the manner in which the breeding animals were selected.

A most striking uniformity in the sex ratios of the inbred litters belonging in the eighth to the twenty-fifth generations of this series is shown by the grouping of the data as made in table 6. The lowest sex ratio (112.5 ♂ : 100 ♀) was found in the first group (eighth to tenth generations); the highest sex ratio (130.7 ♂ : 100 ♀) appeared in the second group (eleventh to thirteenth generations). Between these extremes there was a difference of only 18 points, while in the four following groups of litters the range of variation in the sex ratios was less than 5 points. For the total of 824 inbred litters the sex ratio was 122.3 ♂ : 100 ♀. This latter ratio was not due to an abnormal preponderance of males in a few sets of records, but was based on a series of data that in seventeen out of eighteen cases showed an excess of males greater than that considered as normal for the species. The results obtained, therefore, seem to indicate that by selecting breeding animals from litters that contain an excess of males, the sex ratio can be swung in the direction of the selection, although the line is continually inbred, brother and sister. There was in this case, however, no cumulative effect of the selection. The sex ratios were more uniform in the later generations than in the earlier ones, but they were no higher. It is rather an odd coincidence that the sex ratios in the inbred litters of the eighth and of the twenty-fourth generations were exactly the same (118.2 ♂ : 100 ♀).

Data given in table 4 show that in the half-inbred litters produced in the eighth to the twenty-fifth generations of the A series the range of variation in the sex ratios was from 99 to 140.8 males for each 100 females, six of these ratios being slightly below the norm.   When the data were combined in generation groups (table 6), *it was found that not a single group gave a sex ratio as low as the norm.*   The sex ratios for the litters in the later generation groups were somewhat more uniform than those for the litters in the earlier generation groups, but the uniformity was not as striking as that in the corresponding groups of inbred litters.   For the total of 678 half-inbred litters the sex ratio was 115.6 ♂ : 100 ♀.   This ratio was some 11 points above the norm and less than 7 points lower than the sex ratio in the inbred litters belonging to the same group of generations (122.3 ♂ : 100 ♀). While the litters produced by the mating of inbred females with outbred stock males thus tended to have a lower sex ratio than did the strictly inbred litters, they did not give the sex ratio that was to be expected according to the current view that chance alone determines whether a male-producing or a female-producting spermatozoön shall fertilize the egg.   Such an hypothesis requires that the sexes shall appear in approximately equal numbers when large series of sex data are examined.   In the present case the proportion of the sexes among the 5230 individuals obtained was very far from equal.   In only one group (ninth generation) out of eighteen was there a nearly equal proportion of the sexes, in all other groups there was a pronounced excess of males.

The first twenty-five generations of the A series of inbreds comprised 1752 litters containing 13,116 individuals, 7116 males and 6000 females.   The sex ratio for this series of animals was 117.4 ♂ : 100 ♀.   This ratio was over 12 points above the norm, and since it was based on data for such a large group of animals, it would seem to indicate that in the rat the sex ratio can be altered by selection within a closely inbred line.   In this instance the relative number of males was apparently increased by selecting breeding females from litters that contained an excess of males.

The sex data for the inbred and for the half-inbred litters of the B series, combined in generation groups, are shown in table 7.

TABLE 7

Showing, by generation groups, the sex ratios in the inbred and in the half-inbred litters of the B series (female line)

GENERATION GROUPS	INBRED (FIRST AND SECOND LITTERS)					HALF-INBRED (THIRD AND FOURTH LITTERS)					SUMMARY OF ALL LITTERS				
	Number of litters	Number of individuals	Males	Females	Number of males to 100 females	Number of litters	Number of individuals	Males	Females	Number of males to 100 females	Number of litters	Number of individuals	Males	Females	Number of males to 100 females
1–4	36	233	122	111	109.9	15	109	53	56	94.6	51	342	175	167	104.8
5–7	100	699	364	335	108.7	78	603	300	303	99.0	178	1302	664	638	104.1
1–7	136	932	486	446	109.0	93	712	353	359	98.3	229	1644	839	805	104.2
8–10	104	795	382	413	92.5	73	491	252	239	105.4	177	1286	634	652	97.2
11–13	120	876	400	476	84.0	99	763	358	405	88.4	219	1639	758	881	86.0
14–16	130	·888	386	502	76.9	116	928	450	478	94.1	246	1816	836	980	85.3
17–19	138	1083	472	611	77.3	113	897	414	483	85.7	251	1980	886	1094	80.9
20–22	152	1102	505	597	84.6	109	845	401	444	90.3	261	1947	906	1041	87.0
23–25	150	1149	506	643	78.7	123	875	414	461	89.8	273	2024	920	1104	83.3
8–25	794	5893	2651	3242	81.8 ±1.56	633	4799	2289	2510	91.1 ±1.74	1427	10692	4940	5752	85.9 ±1.39

In the B series, as in the A series, there was a wide range of variation in the sex ratios of the litters produced in the first seven generations (table 5). When the data were combined in generation groups (table 7), the sex ratio in the 136 inbred litters (109 ♂ : 100 ♀) was found to be above the norm, while that in the half-inbred litters (98.3 ♂ : 100 ♀) was below the norm. These two ratios so nearly balance each other that for the total of 229 litters the sex ratio was 104.2 ♂ : 100 ♀, or less than 1 point below the norm: in the corresponding litters of the A series the sex ratio was 8 points above the norm (113.2 ♂ : 100 ♀). On combining the records for the first seven generations of the two inbred series (A, B), it was found that the total of 479 litters gave a sex ratio of 108.6 ♂ : 100 ♀. While this ratio is over 3 points above the norm, it is not sufficiently high to warrant the conclusion that the normal sex ratio was changed through inbreeding, particularly as the ratio was due in great part to an unusual excess of males in the half-inbred litters of the A series (table 4).

As far as can be judged from the results of this part of the investigation, close inbreeding, even when the animals are poorly nourished, does not increase the proportion of male offspring to any extent.

The breeding females in the seventh generation of the B series of inbred were all taken from litters that contained an excess of females; among their offspring the sex ratio was 94.3 ♂ : 100 ♀. *In not one of the subsequent generations was the sex ratio in the inbred litters as high as the norm*, the nearest approach to the norm was in the twelfth generation, where the sex ratio was 101.3 ♂ : 100 ♀ (table 5). In these inbred litters, as in the corresponding ones of the A series, the sex ratios were more uniform in the later than in the earlier generations, but there was no cumulative effect of selection in either case. In the B series, after the thirteenth generation, there was very little change in the relative proportion of the sexes from one generation to the next, and some of the variation found, as stated for the A series, can doubtless be ascribed to environmental action.

When the data for the inbred litters of the eighth to the twenty-fifth generations of the B series were combined in generation groups (table 7), *it was found that the sex ratios for the various groups showed even greater deviations from the norm than did those for corresponding litter groups in the A series, but that this deviation was in the reverse direction, i.e., the number of females born greatly exceeded the number of males.* The highest sex ratio for any group in the B series was 92.5 ♂ : 100 ♀ : for the entire group of 794 litters the sex ratio was 81.8 ♂ : 100 ♀, or 23 points below the norm. This latter ratio is far too low to be considered as a chance variation, and it certainly cannot be attributed to the action of environment. For both series of inbreds were reared simultaneously under the same environmental conditions, and if one ventured to suggest that environment swung the sex ratio in the B series towards the female side it would be necessary to assume that the same environment acted on the animals of the A series in a reverse direction and so swung the sex ratio towards the male side.

As the sex ratio for the inbred litters of the B series was 23 points below the norm, while that for corresponding litters of the A series was 18 points above the norm, it would appear that the sex ratio in the rat can be swung by selection farther towards the female side than towards the male side. Moenkhaus ('11) obtained a similar result in his inbreeding experiments with Drosophila.

The half-inbred litters in the eighth generation of the B series gave a sex ratio nearly 10 points higher than the norm, so here selection was not effective at once in changing the sex ratio. *In none of the subsequent generations, however, was the sex ratio in these litters above the norm, most of them were considerably below it (table 5).* When the data were combined in generation groups (table 7), it was found that the sex ratios for all groups except one (eighth to tenth generations) were very low. For the total of 633 litters the sex ratio was 91.1 ♂ : 100 ♀, thus being 14 points below the norm and 9 points higher than the sex ratio for the inbred litters of this series.

In each of the inbred series the sex ratios in the half-inbred litters belonging in the eighth to the twenty-fifth generations showed less deviation from the norm than did the sex ratios in the corresponding inbred litters, yet in each case the difference between the sex ratio for the inbred group of litters and that for the half-inbred group was less than the difference between the sex ratio for the half-inbred litters and the norm. The possible significance of these results will be discussed later.

In order to obtain the sex ratios for the various generations of the inbred strain as a whole, the data for the two series (A, B) were combined as shown in table 8.

The range of variation in the sex ratios of the litters in the first four generations of the inbred strain was greater than that among all of the other generation groups (table 8). This result was to be expected, considering the relatively small number of individuals in these generations and the adverse conditions under which the animals lived. When the data were combined, however, the sex ratio obtained (110.3 ♂ : 100 ♀) was only 5 points above the

TABLE 8

Showing the sex data for each of the first twenty-five generations of the inbred strain (series A, B), also the sex ratios when the data were combined in generation groups

GENERA-TIONS	NUMBER OF LITTERS	NUMBER OF INDIVIDUALS	MALES	FEMALES	NUMBER OF MALES TO 100 FEMALES	NUMBER OF MALES TO 100 FEMALES IN GENERATION GROUPS
1	5	34	21	13	161.5	
2	20	137	65	72	90.3	
3	36	180	90	90	100.0	
4	68	442	240	202	118.8	110.3
5	132	952	470	482	97.5	
6	107	745	392	353	111.0	
7	111	766	417	349	119.5	108.0
8	103	772	402	370	108.6	
9	129	910	466	444	104.9	
10	141	1034	514	520	98.8	103.6
11	154	1140	594	546	108.8	
12	147	1127	586	541	108.3	
13	159	1229	609	620	98.2	104.8
14	161	1215	619	596	103.9	
15	161	1255	649	606	107.1	
16	159	1155	567	588	96.4	102.5
17	184	1459	728	731	99.6	
18	166	1307	660	647	102.0	
19	166	1313	642	671	95.7	99.1
20	178	1294	671	623	107.7	
21	184	1441	717	724	99.0	
22	190	1502	750	752	99.7	101.8
23	170	1319	663	656	101.1	
24	190	1386	689	697	98.9	
25	187	1338	674	664	101.5	100.4
	3408	25452	12895	12557	102.7 ±1.28	

norm. The sex ratios in the litters of the fifth to the twenty-fifth generations varied from 95.7 to 119.5 males to each 100 females. Variation, it will be noted, was around the norm, eight of the twenty-one ratios being at or above the norm, the rest below it. When combined in generation groups the sex data gave a very uniform series of ratios, as the last column of table 8 shows —not one of these ratios varied more than 6 points from the norm. A variation as great as this would doubtless be found in the sex ratios of any other large series of albino rats, regardless of the manner in which the animals were bred. For the 3256 individuals comprised in the first seven generations of the inbred strain the sex ratio was 108.6 ♂ : 100 ♀. This ratio is sufficiently close to the norm, I think, to indicate that, in the rat, inbreeding per se does not produce a marked increase in the number of male offspring. The sex ratio in the 22,196 individuals in the remaining eighteen generations was 101 ♂ : 100 ♀ : for the entire series of 25,452 animals in the inbred strain the sex ratio was 102.7 ♂ : 100 ♀. While these last two ratios are slightly below the norm, it is evident that in the inbred strain as a whole the sex ratio was not greatly influenced either by inbreeding or by selection. The very different sex ratios obtained in the two series of the inbred strain seem to show, however, that through selection the one inbred strain was separated into two distinct lines, one line (A) having a tendency to produce an excess of males, the other line (B) tending to produce a preponderance of females.

Unfortunately, one cannot predict with certainty what the sex ratio will be in the litters cast by any given inbred female, neither does the sex ratio in the litters cast by one female give a clear indication regarding the proportion of the sexes that will be found among the offspring of a sister rat. It is only by taking the averages for a large number of litters in a given series that the change in the sex ratio is made manifest. As an illustration of the individual differences in females regarding their tendencies to cast young of a certain sex, four sets of data for litters cast by sister females are shown in table 9. In each case given, sister rats were first paired with the same litter brother and later with the same stock male.

TABLE 9

Showing the difference between inbred sisters regarding their tendency to produce an excess of male or of female young when mated with the same male

LITTER SERIES	NUMBER OF YOUNG	MALES	FEMALES	SIRE	LITTER SERIES	NUMBER OF YOUNG	MALES	FEMALES	SIRE
					1				
	$11B_{73}$					$11B_{74}$			
1	11	3	8	$11B_{73}$	1	10	6	4	$11B_{73}$
2	11	5	6	$11B_{73}$	2	11	8	3	$11B_{73}$
3	10	2	8	Stock	3	11	7	4	Stock
					4	9	5	4	Stock
	32	10	22			41	26	16	
					2				
	$17B_{14}$					$17B_{15}$			
1	8	3	5	$17B_{16}$	1	10	4	6	$17B_{16}$
2	9	4	5	$17B_{16}$	2	10	3	7	$17B_{16}$
3	8	3	5	Stock	3	7	3	4	Stock
4	3	1	2	Stock	4	11	5	6	Stock
	28	11	17			38	15	23	
					3				
	$12A_{134}$					$12A_{135}$			
1	8	6	2	$12A_{136}$	1	8	3	5	$12A_{136}$
2	8	4	4	$12A_{136}$	2	9	4	5	$12A_{135}$
3	7		7	Stock	3	5	4	1.	Stock
4	9	3	6	Stock	4	9	5	4	Stock
	32	13	19			31	16	15	
					4				
	$13A^2_{44}$					$13A^2_{45}$			
1	9	5	4	$13A^2_{48}$	1	7	5	2	$13A^2_{48}$
2	10	5	5	$13A^2_{48}$	2	9		9	$13A^2_{48}$
3	10	5	5	Stock	3	12	7	5	Stock
4	10	5	5	Stock	4	8	5	3	Stock
	39	20	19			36	17	19	

The first set of records given in table 9 shows the very great difference in the sex tendencies of two sister rats belonging in the B series. Female 11B₇₃ had cast three litters when she developed pneumonia and had to be killed. Each of these litters contained such a large excess of females that among her thirty-two offspring the sex ratio was only 45.4 ♂ : 100 ♀: Female 11B₇₄, on the other hand, showed a very strong tendency to produce male young, whether she was paired with a brother or with a stock male; among her forty-one offspring the sex ratio was 162.5 ♂ : 100 ♀. As yet no other sister rats have shown such a pronounced difference in their sex tendencies.

A very great similarity in the sex tendencies of sister rats is shown by the second set of records in table 9. Each litter cast by 17B₁₄ and by 17B₁₅ contained an excess of female young, whether the sire of the litter was an inbred or a stock male. In each group of litters the sex ratio was about 65 ♂ : 100 ♀.

Female 12A₁₃₄ produced an excess of male young in each of the two litters sired by her brother, but the two litters sired by a stock male showed a very great excess of female young. Conversely, while female 12A₁₃₅ cast more female than male young when paired with a brother, she showed a strong tendency to produce an excess of male young when mated with a stock male.

The last set of records in table 9 shows a case where the total number of offspring produced by each of two sister rats contained a nearly equal proportion of the sexes, but this proportion was attained in very different ways. Female 13A$_{45}^2$ showed a most pronounced tendency to produce an equal number of male and female young in each of her four litters. In the litters of female 13A$_{46}^2$ the sexes were very unequally distributed; one litter of nine young consisted entirely of females—a most unusual phenomenon in a litter of such size.

Numerous other cases, similar to the ones given, could be furnished from the records for these inbred rats. The cases cited are sufficient, I think, to show the individual differences in the females regarding their tendencies to cast male or female offspring. Incidentally, these records show, also, that the female plays a more important rôle in determining the sex ratio than is generally believed.

An examination of the sex data for successive litters cast by many hundreds of female rats does not indicate that there is "a change of sex tendency from litter to litter" in the female, as Papanicolau ('15) states is the case in guinea-pigs. Such a tendency is not shown in any of the cases given in table 8, and while the sex-proportions in the litters do change in many cases, the change is not general or striking enough to warrant the conclusion that there is a definite sex-determining factor involved.

4. THE SEX RATIOS IN THE LITTERS OBTAINED BY THE MATING OF STOCK FEMALES WITH INBRED MALES

As a check for the results obtained by the mating of inbred females with stock males, series of stock females were bred to males from various generations of the inbred strain. The number of such experiments was small, considering the scale on which the main series of experiments was conducted, but the results obtained were uniform enough to be significant.

The stock females used in these experiments were reared under the same environmental conditions as the inbred rats. When they were about three months old they were paired with males from the A or from the B series that had sired inbred litters. In order to make this series of records more strictly comparable to that obtained in the inbred strain, only four litters from any one female were recorded.

The data for the litters obtained by the mating of stock females with inbred males are given in table 10.

Stock females paired with males from the fifth generation of the inbred strain produced litters in which the sex ratio was below the norm, whether the sire of the litter belonged to the A or to the B series of inbreds. The litters sired by males from A series, however, had a much higher sex ratio than did those sired by males from the B series, although at the fifth generation there was no selection of breeding animals in the inbred strain according to a definite plan. The eighteen litters in this series gave a sex ratio of 94.7 ♂ : 100 ♀, or 10 points below the norm. This ratio might seem to indicate that inbred males tended to produce an excess of female offspring, but the number of litters

TABLE 10

Showing the sex ratios in litters produced by the mating of outbred stock females with inbred males

INBRED SERIES TO WHICH SIRES BELONGED	GENERATION TO WHICH SIRES BELONGED	NUMBER OF LITTERS	NUMBER OF INDIVIDUALS	MALES	FEMALES	NUMBER OF MALES TO 100 FEMALES
A	5	10	59	30	29	103.5
B	5	8	52	24	28	85.7
		18	111	54	57	94.7 ± 4.30
A	9	7	51	28	23	121.7
A	12	12	60	30	30	100.0
A	13	5	33	20	13	153.9
A	15	19	177	79	98	80.6
A	16	11	126	67	59	113.6
A	17	14	117	60	57	105.2
A	18	39	347	177	170	104.1
		107	911	461	450	102.3 ± 5.88
B	10	12	97	51	46	110.9
B	12	11	75	38	37	102.7
B	13	8	42	18	24	75.0
B	15	19	172	87	85	102.3
B	16	29	243	116	127	91.3
B	18	38	313	152	161	94.4
		117	942	462	480	96.2 ± 3.14
Total...	....	242	1964	977	987	99.1

3.14

examined was too small to warrant any general conclusion from the results obtained.

The second section of table 10 shows the sex ratios in the various groups of litters obtained from the matings of stock females with males from the ninth to the eighteenth generations of the A series of inbreds. There was considerable variation in these sex ratios, as was to be expected considering the number of animals involved. The total of 107 litters in this group gave a sex ratio of 102.3 ♂ : 100 ♀. This ratio, it will be noted, was below the norm, although the sires of the litters were males that, paired with their sister, had fathered litters in which there was, as a rule, a preponderance of male young.

The sex ratios in the various groups of litters obtained by the mating of stock females with males from the tenth to the eighteenth generations of the B series of inbreds showed a much narrower range of variation than that found in the litters sired by males of the A series of inbreds, although the number of litters produced in the two series was about the same. The 117 litters in this group gave a sex ratio of 96.2 ♂ : 100 ♀, which was 9 points below the norm. Any significance that this ratio might seem to have, when taken alone, is apparently annulled by the fact that the sex ratios for the other litters groups were also below the norm, whether the sires of the litters came from the A or from the B series of inbreds. Moreover, the probable error of the mean, calculated from the averages for the various sets of litters, was so large in every case that the differences between the sex ratios of the various groups were rendered valueless.

The 242 litters in this series gave a sex ratio of 99.1 ♂ : 100 ♀. While this ratio was some 6 points below the norm, it differed by only 4.4 points from the sex ratio found in the 1510 litters obtained by the mating of inbred females with stock males (103.5 ♂ : 100 ♀). The results as a whole, therefore, do not indicate that the sex ratio was influenced to any extent by the fact that the sires of the litters were inbred rather than outbred males.

The final experiment to be made, the pairing of females from the one inbred series with males from the other inbred series, was not begun until the animals reached the twenty-sixth generation. The number of litters as yet obtained is too small to afford a basis for any general conclusion, but thus far females of the A series (male line) when paired with males from the B series (female line) have produced more male than female young, and, conversely, females of the B series, when paired with males of the A series, have shown a tendency to cast more female than male young.

The results of these various series of experiments are summarized and discussed in the following section.

5. DISCUSSION

As a basis for discussion the results obtained in this investigation are briefly summarized as follows:

1. The inbreeding of litter brother and sister for six consecutive generations. during which there was no selection of animals for breeding, did not increase the number of male offspring to any extent. The sex ratio in the 3256 young obtained was 108.6 ♂ : 100 ♀, or less than 4 points above the sex ratio taken as the norm (105 ♂ : 100 ♀).

2. Beginning with the seventh generation all breeding females in the A series were taken from litters that contained an excess of males. After this time the females in this series tended to produce an excess of male young, whether they were paired with a litter brother or with an unrelated stock male (table 6). The litters sired by inbred males gave a sex ratio of 122.3 ♂ : 100 ♀, or over 17 points above the norm; while the litters sired by stock males showed a sex ratio of 115.6 ♂ : 100 ♀, or nearly 11 points above the norm.

3. From the time that the breeding females in the B series were selected from litters containing an excess of females (seventh generation), the litters produced showed a reverse proportion of the sexes to that shown by corresponding litters in the A series (table 7). Litters sired by inbred males had a very low sex ratio (81.8 ♂ : 100 ♀); in the litters sired by stock males the sex ratio was 9 points higher than that in the inbred litters (91.1 ♂ : 100 ♀), but it was still significantly lower than the norm.

4. On combining the data for the two inbred series it was found that among the 25,452 individuals comprised in the inbred strain the sex ratio was 102.7 ♂ : 100 ♀, or less than 3 points below the norm (105.0 ♂ : 100 ♀). It thus appears that through selection the inbred strain was separated into two distinct lines: one (A) showing a high sex ratio, the other (B) a low sex ratio. Selection had the greater influence on the female line, since the sex ratio for the litters of the B series showed greater deviation from the norm than did that for the litters of the A series.

5. Stock females mated with inbred males tended to produce litters in which the sex ratio was below the norm, regardless of

whether the male belonged to the A or to the B series of inbreds. The litters sired by males from the A series showed a
higher sex ratio (102.3 ♂ : 100 ♀), however, than did the litters
sired by males from the B series (96.2 ♂ : 100 ♀), but these ratios
are not significant, since they differ from each other and from the
norm by less than three times the probable error (table 10).

Düsing's contention that close inbreeding increases the relative
number of male offspring was based mainly on statistics of human
births collected from several isolated communities in which there
were many consanguineous marriages, and on the supposedly great
preponderance of male births among the Jews, who are a clannish
race and intermarry more frequently than do other civilized
races. The latter evidence is undoubtedly invalid, as Pearl and
Salaman ('13) have shown that the normal sex ratio among the
Jews is the same as that in other races of man (105 ♂ : 100 ♀),
and that the anomalous sex ratio among them is due to faulty
registration, male births being recorded where those of females are
not. The high sex ratio in the other cases cited by Düsing can
doubtless be attributed to a similar cause. The great excess of
males found in various strains of thoroughbred dogs Heape ('08)
ascribed in part to inbreeding, but in these cases also it is probable that the statistics are not reliable, since female pups are
commonly discarded from large litters and males are registered
more often, as a rule, than females.

The inbreeding experiments of Huth ('87) on the rabbit, of
Schultze ('03) and of Copeman and Parsons ('04) on mice were
made with relatively small numbers of animals, and the sex ratios
obtained showed no greater deviations from the norm than might
have been expected under the conditions of the experiments.
Shull ('13) found no change in the sex ratio of Hydatina senta as a
result of repeated inbreeding, the proportion of male-producers
and of female-producers remaining practically constant. In the
present series of inbreeding experiments with the albino rat, all of
the animals belonging to the earlier generations suffered severely
from malnutrition, which Düsing ('84) considered as a very potent
factor in increasing the number of male offspring, yet among the
individuals in the first seven generations the sex ratio was only

slightly higher than the norm.   The results of these various series
of experiments would seem to indicate that inbreeding per se has
little, if any, effect on the sex ratio.

Moenkhaus' ('11) extensive series of inbreeding experiments on
Drosophila so closely parallel my own experiments on the rat,
both in the manner in which the experiments were conducted and
in the results obtained, that a brief résumé of his work must be
given here.

In order to obtain the normal sex ratio in Drosophila, Moenkhaus ascertained the sex of 26,933 imagos that developed from
eggs laid by wild flies, and found among them a sex ratio of 88.8
♂ : 100 ♀.   In this species, therefore, there is normally an excess
of females, as other investigators (Rawls, '13; Hyde, '14; Warren,
'18) have noted.   The experiments were conducted in the following way: "Two pairs were selected from nature, the one
showing a high, the other a low female ratio.   These were selected as the parents of the two strains to be developed.   From
among the offspring of each of these two pairs a number of single
matings were made.   From among these the pair that showed the
most favorable ratio in the desired direction was selected to continue the strain.   The same process was repeated as often as
desired."

In this way Moenkhaus developed two inbred strains in one of
which the individuals showed a high sex ratio, in the other a low
sex ratio.   The results of this part of the investigation showed that
"it is possible to develop a strain with a high female ratio much
more easily and pronouncedly than a male strain."   Moenkhaus
then made reciprocal crosses between the two strains in order to
determine, "first, whether the maternal or the paternal elements
had an equal share in the control of this ratio, and second,
whether this ratio was determined in the process of fertilization."
The experiments showed, in a most decided way, that "the male
has little or no influence in determining the sex ratio in this species.
In most of the cases the ratio of the offspring falls pretty closely
around that of the strain from which the females were taken.
. . . It is not certain, however, that the sex ratio is established
before fertilization, since the experiments do not with certainty

entirely exclude the male influence." In his summary Moenkhaus states: "The sex ratio is one of the qualities that is, like color, an inherent character of this creature, strongly transmissible and amenable to the process of selection. . . . Sex is probably very little, if at all, influenced at fertilization in this species, but it is probably determined much earlier and by the female."[1] Moenkhaus' conclusions regarding the character of the sex ratio and its amenability to selection are as applicable to the rat as they are to Drosophila, judging from the results of my inbreeding experiments on the former species. Neither of these investigations, however, give any information regarding the causes that condition sex, although each seems to indicate that the female takes quite as important a part in this process as does the male.

In the inbred strain, after the animals for breeding were selected in each generation according to a definite plan, the two series (A and B) became two separate lines as regards the sex-proportions among the young. In the one line (A) the litters contained, as a rule, an excess of males; in the other line (B) there was a corresponding excess of females. Between these two lines there was no very marked difference as regards the size of the individuals at a given age, their fertility or longevity, as the data given in previous papers have shown (King, '18, '18 a). Generation after generation, as far as the experiments have been carried, the sex ratios in the inbred lines have remained distinct, and the variations from the norm have been in the same direction in each generation of each series. These results are definite enough, and they are based on data from a sufficiently large number of animals. I think, to warrant the conclusion that in the rat the sex ratio is to a certain extent at least, a character that is amenable to selection.

[1] Warren ('18) has recently repeated Moenkhaus' selection experiments on Drosophila, and concludes that the sex ratio in this form is "not readily, if at all, modifiable by selection." Warren believes that the modified sex ratios found in two of his three series of experiments were due to 'chance variation,' and he attributes the anomalous sex ratios obtained by Moenkhaus to the action of a sex-linked lethal factor—the explanation offered by Morgan ('14 a) to account for the unusually low sex ratios found in several strains of Drosophila.

As the rats had been inbred brother and sister only, they were, according to Fish's ('14) calculations, 79.687 per cent homozygous at the time that the selection of breeding animals began (seventh generation). Selection, if effective at all in changing the sex ratio, should act in one or two generations, unless a considerable number of factors were involved. In the latter case selection might produce a gradual change in the sex ratio which would reach its culmination only after a number of generations. In each series, as table 4 and table 5 show, the sex ratio in the inbred litters of the eighth generation was close to the sex ratio that was the average for all of the litters produced in the eighth to the twenty-fifth generations, and the sex ratios in the later generations showed no greater deviation from the norm than did those in the earlier generations, although they were somewhat more uniform. Selection thus produced its maximum effect at once, and could not shift the centre of gravity of the variation in the direction of the selection, as it did in the experiments which Castle and Phillips ('14) made with piebald rats. It would appear from these results that very few heritable factors concerned in the production of the sex ratio, possibly not more than a single pair, were acted upon by selection, and that, as Pearl ('17) has stated: "selection acts only as a mechanical sorter of existing diversities in the germ plasm and not as a cause of alteration in it."

As sister rats show such marked individual difference regarding their sex tendencies, and as both nutritive (Slonaker and Card, '18) and environmental conditions (King and Stotsenberg, '15) seem to influence the sex ratio in the rat, it would seem that the sex ratio may be modified by so many agencies that it would be useless to attempt to determine the number or the nature of the particular factors that were acted upon by selection in the present case. The factors involved are evidently not of very great potency, and their action is clearly shown only when a relatively large number of animals are closely inbred under environmental and nutritive conditions that are as uniform as it is possible to make them. Whatever their nature, or in whatever manner they may be inherited, I believe that these factors act on the metabolism of the ova in such a way as to render the ova more easily fertilized by one kind of spermatozoa than by the other. In the A series of inbreds, under the conditions given, the ova tended to attract spermatozoa that were 'male-producing;' in the B series, the ova tended to attract spermatozoa that were 'female-producing.'

In advocating the possibility that fertilization may be selective I am aware that I run counter to the general belief that any egg . is capable of fertilization by any spermatozoön that happens to come in contact with it, and that those whose views have much weight in molding biological opinion believe that this hypothesis is "so improbable as almost to invalidate any interpretation into which it enters" (Wilson, '10). Just why this hypothesis is considered so untenable is not clear. It is true that it has not been definitely proved in any case, but neither has it been disproved, nor has any convincing proof been offered, as yet, for the very elaborate hypotheses that have been advanced to account for heredity in general and in specific cases. The burden of proof rests equally upon those who object to this hypothesis as on those who maintain it.

We owe to McClung ('02) the suggestion that the accessory chromosome may be a sex-determinant. In discussing the possible action of the accessory chromosome in determining sex, McClung ('02 a) states: 'even up to the time of fertilization the female elements are so placed as to react readily to stimuli from the mother. Here they are approached by the wandering male elements from which they may choose—if we may use such a term for what is probably chemical attraction—either the spermatozoa containing the accessory chromosome or those from which it is absent. In the female element, therefore, as in the female organism, resides the power to select that which is for the best interest of the species."

In advocating selective fertilization as the probable cause of anomalous sex ratios, Heape ('09) says: "it must be remembered that there are an enormous number of spermatozoa available for the fertilization of each ovum, and, moreover, it will be recollected there are undoubtedly chemotactic properties associated with ova which insure that ova of different species floating in the sea shall each be fertilized by spermatozoa of the same species, so that to grant there is still more delicate chemotaxis at work is not an illegitimate but is indeed a reasonable supposition." Castle ('03) also postulated selective fertilization in the elaboration of his Mendelian theory of sex-determination.

The one attempt that has been made to test the hypothesis of selective fertilization (Morgan, Payne and Browne, '10) seemed to indicate that the egg is fertilized by the first spermatozoön that strikes it 'head-on,' but the conditions under which the observations were made were so abnormal that no definite conclusions from them were possible, and even Morgan ('11) states that the evidence is 'admittedly insufficient.'

An earlier experiment that has a bearing in this connection seems to have been overlooked and therefore needs to be noted here. Marshall ('10) injected into the vagina of a pure-bred Dandie Dinmont bitch a mixture of seminal fluid taken from a pure-bred dog of the same species and from a mongrel terrier of unknown ancestry. Fifty-nine days later the bitch littered, producing four pups which were much alike. One of the pups died early, but as the other three developed into mongrels which resembled the terrier sire there was little doubt but that all four puppies were mongrels. Marshall cites another case in which a Dandie Dinmont bitch copulated with a dog of the same breed and two days later with a Scotch terrier. The bitch littered three pups; one was pure Dandie Dinmont, the other two half-bred Scotch terriers. These cases, according to Marshall, were indicative of a 'selective' on the part of the ova of the pure-bred bitch to "conjugate with dissimilar rather than with related spermatozoa."

I have recently been making a series of experiments somewhat on the order of those cited by Marshall, and the results obtained indicate a very strong tendency on the part of the ova of the albino rat to conjugate with spermatozoa from the wild gray rat rather than with the spermatozoa of the albino rat, although under the conditions of the experiment, details of which will be published later, the advantage in every case was with the spermatozoa from the albino male. If fertilization can be selective in such cases I can see no valid objection to the assumption that the chemotactic reaction between ova and spermatozoa may be even more delicate and thus, under given conditions, make possible the fertilization of an egg by a spermatozoön that has one sex potency rather than the other.

There is another possible interpretation of the anomalous sex ratios found in the inbred litters of the two series. We might assume that inbreeding had acted on the males in some way so as to render one kind of spermatozoön more potent than the other in fertilizing the ova, and that this difference in potency came to have an heritable basis in the germ plasm and so could be acted upon by selection. In the A series of inbreds, according to this assumption, the 'male-producing' spermatozoa became the more potent; in the B series the 'female-producing' spermatozoa came to have the greater potency. Were this assumption correct it should receive confirmation both from the results of the experiments in which inbred females were paired with stock males and from the experiments in which stock females were paired with males from different generations of the two inbred series. The litters obtained in the former case should show a nearly equal proportion of the sexes (provided it was merely a matter of chance which kind of spermatozoa fertilized the ova), since the males were outbred and therefore, theoretically, the two kinds of spermatozoa had equal power to fertilize the ova. In the latter case the litters obtained should show a high sex ratio when the sire came from the A series of inbreds and a low sex ratio when the sire belonged to the B series.

As shown in table 4 and in table 5, the half-inbred litters produced by the mating of inbred females with stock males gave sex ratios that were very far from equality. In only one generation of each series was there an approximately equal proportion of the sexes, in all other cases the variation was in a definite direction: in the A series there was an excess of males; in the B series the females predominated. In both series, moreover, the sex ratios in the half-inbred litters were much closer to those in the corresponding inbred litters than they were to the norm. The uniformity in the various series of records and the small size of the probable error of the mean exclude the possibility that the sex ratios could have been produced by chance or by environmental action. The results, therefore, do not support the contention that the male is the chief factor in determining the sex ratio in the rat.

The sex ratio in each of the three groups of litters obtained by the mating of stock females with males from various generations of the two inbred series was below the norm, whether the sire of of the litters belonged in the A or the B series. The sex ratio in the group of litters sired by males from the A series (102.3 ♂ : 100 ♀) was only 6 points higher than that in the litters sired by males from the B series (96.2 ♂ : 100 ♀). The results in this case, therefore, do not indicate that inbreeding, with selection, influence the potency of the spermatozoa in any way; they seem rather to signify that the particular stock females used for breeding tended to attract spermatozoa that were 'female-producing' rather than those that were 'male-producing.'

The results of the various experiments in which inbred and outbred animals were paired, taken in connection with those from the experiments in which matings were made between litter brother and sister, seem to show that in the rat, as in Drosophila (Moenkhaus), the female has a greater influence than the male in determining the sex ratio, and that chance alone cannot be the factor that determines whether an egg shall be fertilized by a 'male-producing' or by a 'female-producing' spermatozoön.

The size of the probable error of the mean (tables 6 and 7) indicates that in each series the difference between the sex ratio for the group of inbred litters and that for the group of half-inbred litters is a significant one. Apparently, therefore, the chemotactic reaction between the ovum and the spermatozoön is not quite the same where these sexual elements come from unrelated individuals as when they are produced by individuals that are closely inbred. A somewhat analogous case is found in the hermaphroditic ascidian, Ciona, where normally, as Castle ('96) and Morgan ('04, '05) have shown, the eggs are not fertilized by spermatozoa from the same individual, although they are readily fertilized by spermatozoa from any other individual, while the spermatozoa from the first animal are functional when used with ova of another animal. Morgan ('14) has suggested that the infertility of the eggs of Ciona to spermatozoa from the same individual may be due to the similarity in the hereditary complex of the germ cells which in some way decreases the chances of their uniting. The selective fertilization experiments made by Marshall ('10) with different varieties of dogs and also my own experiments with different varieties of rats show that the ova of these animals have a strong tendency to unite with spermatozoa from individuals belonging to unrelated stock rather than with spermatozoa from individuals of the same 'blood.' When my own experiments are completed the results will show, I hope, whether there is a still more delicate chematactic reaction between the ova and the spermatozoa which will lead to the production of more males than females among the hybrid offspring. The anomalous sex ratios that appear in F, hybrids almost invariably show an excess of males. This suggests that the greater the difference between individuals as regards theis blood relationship the stronger is the attraction between the ova and the 'male-producing spermatozoa. If this suggestion proves true, its converse ought also be to true, and in a closely inbred line we would expect that the chemotactic reaction between the ova and the spermatozoa would be such that an excess of females would be produced. Such a possibility is not incompatible with the results of the present investigation, since in the inbred strain, as a whole, the sex ratio was below the norm, while the sex ratios in the litters of the female line (B) showed a greater deviation from the norm than did the sex ratios in the litters of the male line (A).

The results of this series of experiments, as a whole, seem to indicate that in the rat, as in the pigeon (Riddle, '14, '16, '17), in Drosophila (Moenkhaus, '11) and in the guinea-pig (Papanicolau, '15), the female has more influence in determining the sex ratio than has the male. Yet it is not in the differentiation of the ova, nor in the development of the spermatozoa, that the key to the riddle of sex-determination will be found. A knowledge of the interaction of the germ cells, and of the conditions that influence it, must be gained before the final solution of this problem can be attained.

STUDIES ON INBREEDING

IV. A FURTHER STUDY OF THE EFFECTS OF INBREEDING ON THE GROWTH AND VARIABILITY IN THE BODY WEIGHT OF THE ALBINO RAT

HELEN DEAN KING

The Wistar Institute of Anatomy and Biology

EIGHT CHARTS

In order to complete the series of records for the first twenty-five generations of inbred albino rats, data showing the growth and variability in the body weights of individuals belonging in the sixteenth to the twenty-fifth generations are given in the present paper.

Five litters from each generation of the two inbred series (A and B), comprising a total of 296 males and 310 females, were used for this study. The rats in these litters were selected in the same manner, and they were weighed at the same age periods, as were the individuals of the seventh to the fifteenth generations for which body-weight records were taken (King, '18). The data for the animals in the different generations of the inbred strain are therefore strictly comparable.

During the past three years, when most of the weighings were taken, it was not possible to rear the animals under environmental and nutritive conditions that were as favorable to growth and to fertility as those existing previously. Owing to economic conditions incident to the war, it became necessary to make a radical change in the character of the food that the rats received. The 'scrap' food (carefully sorted table refuse), on which the animals of the earlier generations seemed to thrive exceedingly well, had to be replaced by a ration that consisted, for the most part, of oats and corn, with the occasional addition of various kinds of vegetables and a little meat. Some of the available

135

substitutes that from time to time were added to the diet in order to vary it, such as alfalfa, linseed and cottonseed meal, proved very injurious to the rats and very materially affected their growth and fertility. For some time, therefore, the food given the animals has been largely in the nature of an experiment, and it has not even yet been possible to work out a ration that produces as rapid and vigorous growth and that is as favorable to reproduction as was the 'scrap' food given previously.

Extremes of temperature, either of heat or of cold, have a very marked effect on the body growth of the rat, as they have on that of mice (Sumner, '09), and many of the animals in the later generations of the inbred strain suffered considerably from this cause. During the excessive cold of the winter of 1917–1918 it was impossible to keep the colony house above the freezing point for days at a time, and in consequence the rats ceased growing at a normal rate and many of them developed pneumonia. The periods of intense heat experienced during the summer of 1918 also had a very deleterious effect on the vitality and on the body growth of the rats. As a result of the combined action of these various factors, all inimical to growth as well as to reproduction, the rats of the eighteenth to the twenty-fifth generations were severely handicapped, and they did not increase in body weight as rapidly, nor did they attain as great a maximum body weight, as did the individuals of the earlier generations. That this decrease in the size of the inbred animals was caused by unfavorable conditions of environment and of nutrition, and not by continued inbreeding, is shown conclusively by the fact that the body weights of hundreds of rats in the outbred-stock colony were just as seriously affected by these adverse conditions as were those of the inbred rats, as will be shown later.

Data showing the average body weights at different ages of 179 males and of 130 females belonging in the sixteenth to the twenty-fifth generations of the A series of inbred rats are given in table 1 and in table 2: similar data for 117 males and for 180 females belonging in the same generations of the B series of inbreds are given in table 3 and in table 4.

TABLE 1

Showing, by generations, the average body weights at different ages of 179 males belonging in the sixteenth to the twenty-fifth generations of the A series of inbred rats

AGE	GENERATIONS									
	16	17	18	19	20	21	22	23	24	25
days										
13	19	18	19	18	17	20	19	17	16	18
30	44	44	49	41	42	45	46	44	39	43
60	131	121	135	97	118	115	111	96	83	104
90	188	186	192	142	186	164	165	126	121	163
120	232	228	223	197	237	211	203	169	159	200
151	255	253	259	232	259	244	226	207	188	233
182	274	268	286	262	280	271	250	231	216	252
212	296	277	309	286	289	295	276	243	231	277
243	302	288	328	298	297	311	291	254	246	279
273	321	310	352	298	308	310	308	264	265	292
304	317	321	356	301	309	305	305	279	272	303
334	327	322	365	305	313	311	313	290	276	316
365	333	319	378	308	324	322	320	288	284	325
395	336	332	394	304	334	318	306	298	287	327
425	331	339	376	295	340	319	293	297	290	332
455	320	332	361	295	357	316	296	289	293	322
Number rats weighed	17	17	14	17	15	16	20	21	21	21

TABLE 2

Showing, by generations, the average body weights at different ages of 130 females belonging in the sixteenth to the twenty-fifth generations of the A series of inbred rats

AGE	GENERATIONS									
	16	17	18	19	20	21	22	23	24	25
days										
13	19	17	19	17	16	19	17	16	16	17
30	41	44	48	39	40	44	43	42	38	41
60	99	105	110	94	97	95	99	81	79	94
90	141	163	148	134	153	135	142	107	109	131
120	170	178	179	163	177	166	170	135	137	153
151	187	197	195	182	195	192	179	163	160	177
182	206	211	208	197	205	203	183	184	178	180
212	210	214	219	205	215	209	202	190	185	191
243	214	225	222	212	221	213	207	194	187	194
273	230	221	224	214	222	218	208	197	190	204
304	225	233	224	221	216	230	206	206	199	212
334	229	232	220	225	215	236	215	209	189	215
365	240	231	221	223	214	242	217	211	194	223
395	242	227	224	219	210	239	217	209	195	234
425	235	231	224	214	216	231	216	208	190	240
455	243	236	223	216	212	229	215	204	186	230
Number rats weighed	13	11	12	13	13	14	14	13	13	14

Tables 1 to 4 are inserted mainly for reference, but a comparison of the data for the males and females in the various generations brings out clearly the relation between the two sexes as regards their relative body weights at different age periods. In some few instances the average body weights of the males and of the females in a given generation were the same when the animals were thirteen or thirty days old, but after this age the males were the heavier at each period for which records were taken. A similar relation between the body weights of the sexes was also noted for the inbred animals of the seventh to the fifteenth generations (King, '18; tables 1 to 4). Investigations in which large series of stock Albinos were weighed at stated periods (Donaldson, '06; Jackson, '13; King, '15; Hoskins, '16) have shown likewise that, with few exceptions, the average body weight of the males exceeds that of the females at each weighing period. Since the data for all generations of the inbred strain is in full accord with that for various series of stock Albinos, it is evident that inbreeding through twenty-five generations of brother and sister matings has not changed the normal relative body weights of the sexes at any age period for which records have been taken.

For the purpose of analysis and to facilitate a comparison between the growth in body weight of the individuals in the later generations of the inbred series with those in the earlier generations, the body-weight data for the animals belonging in the sixteenth to the twenty-fourth generations of each inbred series were combined in groups of three generations each: the data thus combined are shown in tables 5 to 7. In each of these tables the data for the individuals of the twenty-fifth generation are given separately in order to show the status of the animals at the end of this period of inbreeding.

Data indicating the growth in body weight of males and of females belonging in the various generation groups of the A series of inbreds are shown in table 5.

As a graphic representation of series of data greatly facilitates their comparison, the body-weight data for various groups of albino rats, given in tables 5 to 11, have formed the basis

TABLE 3

Showing, by generations, the average body weights at different ages of 117 males belonging in the sixteenth to the twenty-fifth generations of the B series of inbred rats

AGE	GENERATIONS									
	16	17	18	19	20	21	22	23	24	25
days										
13	19	17	20	21	18	19	20	19	18	18
30	52	44	51	48	41	43	48	48	42	42
60	119	111	119	123	109	111	96	114	95	114
90	178	165	181	175	161	168	159	154	143	147
120	213	222	225	197	203	220	188	180	182	180
151	237	245	261	227	233	245	229	212	224	221
182	266	266	284	251	263	267	263	235	228	231
212	284	279	297	277	272	277	279	252	240	252
243	289	289	310	304	279	305	284	258	259	279
273	296	297	319	323	285	312	293	265	272	289
304	304	303	323	324	289	317	300	276	271	299
334	310	319	336	337	298	323	301	291	292	312
365	322	317	322	337	295	328	315	290	305	322
395	337	326	311	352	296	322	309	293	303	339
425	334	354	305	364	290	320	305	300	298	352
455	340	299	299	350	279	322	299	297	287	349
Number rats weighed	12	11	11	11	11	12	13	12	12	12

TABLE 4

Showing, by generations, the average body weight at different ages of 180 females belonging in the sixteenth to the twenty-fifth generations of the B series of inbred rats

AGE	GENERATIONS									
	16	17	18	19	20	21	22	23	24	25
days										
13	19	16	17	20	17	18	19	19	17	18
30	47	41	48	44	39	41	44	45	41	41
60	102	93	107	101	92	101	85	97	83	93
90	151	125	157	143	133	149	133	133	119	127
120	166	165	179	167	162	192	151	149	143	152
151	189	180	199	183	179	188	177	166	167	166
182	199	194	212	194	198	199	194	177	177	178
212	211	205	214	205	201	200	199	187	185	187
243	215	214	216	212	216	210	199	196	191	191
273	221	226	218	216	220	210	203	196	194	194
304	224	217	217	221	218	211	200	198	196	202
334	225	224	215	225	217	209	207	208	203	204
365	233	224	220	232	214	212	205	208	211	214
395	239	216	216	232	212	210	207	212	201	213
425	243	238	218	233	207	203	203	210	199	220
455	243	253	217	228	201	202	202	211	200	218
Number rats weighed	16	14	18	19	19	18	18	19	19	20

139

for the construction of the graphs shown in figures 1 to 8. The graphs in figure 1 show the growth in body weight of four generation groups of male rats belonging in the A series of inbreds (data in table 5). In this, as in some of the other figures, the graphs should properly run very close together or overlap in various places. If, however, the graphs had been drawn in this manner, it would be difficult to follow their course, and therefore

TABLE 5

Showing the average body weights at different ages of inbred rats of the A series, separated into groups according to the generation to which the individuals belonged

AGE	MALES				FEMALES			
	Genera-tions 16–18	Genera-tions 19–21	Genera-tions 22–24	Genera-tion 25	Genera-tions 16–18	Genera-tions 19–21	Genera-tions 22–24	Genera-tion 25
days								
13	19	18	17	18	18	17	16	17
30	46	43	42	43	44	41	41	41
60	129	110	96	103	105	95	86	94
90	189	163	137	163	150	139	120	131
120	228	214	177	200	175	169	144	153
151	255	244	206	233	192	190	168	172
182	275	271	231	252	208	202	185	180
212	291	290	249	277	214	209	193	191
243	301	302	264	279	220	215	196	194
273	322	306	280	292	225	218	198	204
304	326	305	286	303	227	221	204	212
334	333	310	294	316	227	222	204	215
365	337	317	298	325	232	224	208	223
395	340	316	299	327	251	220	208	234
425	345	313	294	331	229	218	205	240
455	338	316	293	322	231	218	202	230

the space between them has been arbitrarily widened in some places in order to keep the lines distinct.

While the general course of all of the graphs in figure 1 is much the same, their relative position clearly shows the progressive decrease in body weight that has resulted from the action of unfavorable conditions of environment and of nutrition. The rats in the sixteenth to the eighteenth generations were fed, for the most part, on 'scrap' food, and, as graph A in figure 1 shows, the males of the A series that belonged to these generations were heavier at all ages than were the males in the later generation groups, excepting at the 243-day period. Rats in the nineteenth to the twenty-first generations were not greatly affected by the change in diet, as for some months it was possible to give them 'scrap' food part of the time. The males of this generation

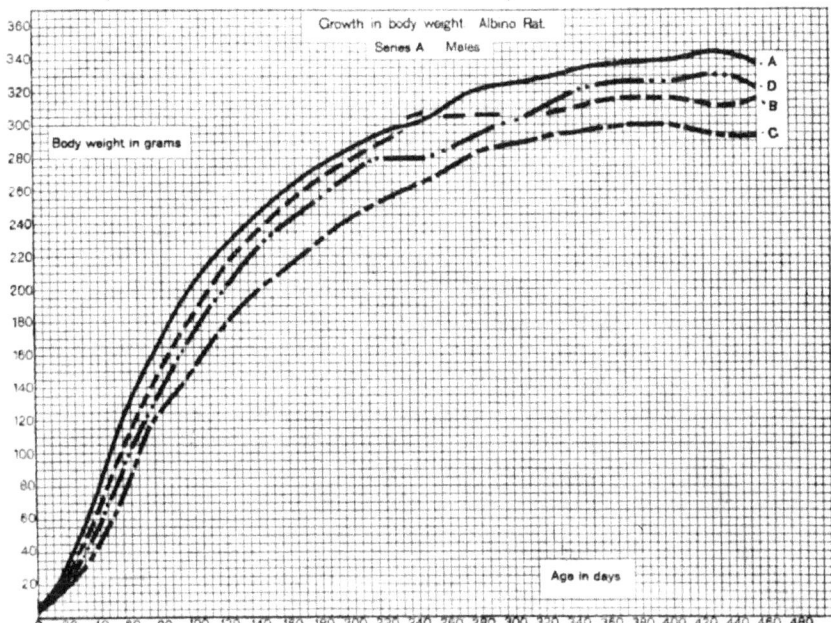

FIG. 1 Graphs showing the increase in the weight of the body with age for males belonging to various generation groups of the A series of inbred rats. A, graph for males of the sixteenth to the eighteenth generations, inclusive; B, graph for males of the nineteenth to the twenty-first generations, inclusive; C, graph for males of the twenty-second to the twenty-fourth generations, inclusive; D, graph for males of the twenty-fifth generation (data in table 5).

group, as the position of graph B indicates, were nearly as large as were those of the earlier generation group during the adolescent period, but in the adult state their body weights fell off rapidly. Individuals in the twenty-second to the twenty-fifth generations of the inbred strain suffered most severely from the altered food conditions as well as from extremes of temperature, and the males of the A series were very inferior in body weight to those of the preceding generations, as graph C and graph D in figure 1 show. Since the number of weighed individuals in a single generation was comparatively small, it is not surprising that the course of graph D should be rather erratic. At its beginning this graph runs very slightly higher than graph C, but at the 90-day period it begins to rise rapidly, and at 334 days it crosses graph B and subsequently runs above it until the final weighing. In the A series of inbreds the males of the

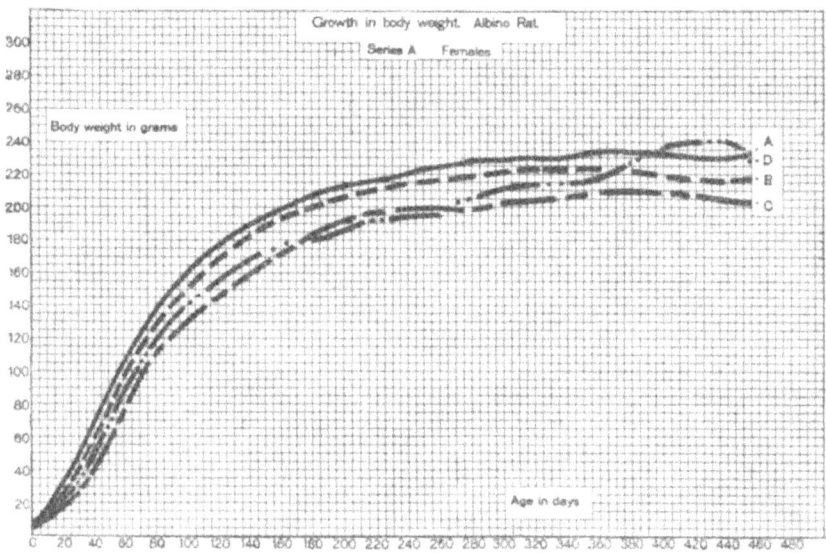

· Fig. 2 Graphs showing the increase in the weight of the body with age for females belonging to various generation groups of the A series of inbred rats (data in table 5; lettering as in fig. 1).

twenty-fifth generation were, as a group, superior in body weight to the males of the generation preceding. The superiority of these individuals can be attributed in part to an improvement in the nutritive conditions and in part to the fact that the majority of animals in this generation were born at the time of year that experience has shown is most favorable for body growth in the rat, i.e., the winter months.

Graphs showing the growth in body weight of females belonging to various generation groups of the A series of inbreds are shown in figure 2. The data from which these graphs were constructed are given in table 5.

In general the relative position of the graphs in figure 2 is much the same as that of the graphs in figure 1. Graph A, representing the body weight increase with age for females of the sixteenth to the eighteenth generations, runs higher than any of the other graphs for the greater part of its course, while the position of the other graphs indicates that there was a gradual decrease in

TABLE 6

Showing the average body weights at different ages of inbred rats of the B series, separated into groups according to the generation to which the individuals belonged

AGE	MALES				FEMALES			
	Genera-tions 16–18	Genera-tions 19–21	Genera-tions 22–24	Genera-tion 25	Genera-tions 16–18	Genera-tions 19–21	Genera-tions 22–24	Genera-tion 25
days								
13	19	19	19	18	17	18	18	18
30	49	44	46	43	46	42	43	41
60	116	115	102	114	102	98	89	93
90	175	168	152	147	146	141	128	127
120	220	207	184	180	170	173	148	152
151	249	235	216	211	190·	183	170	166
182	271	260	243	231	202	197	183 ·	178
212	286	275	258	252	211	202	190	187
243	295	297	268	279	215	213	195	191
273	303	308	277	289	221	215	198	194
304	309	311	283	299	220	216	198	202
334	319	319	296	312	222	217	206	203
365	320	319	304	322	227	219	207	210
395	326	319	301	339	230	217	207	213
425	328	323	301	352	234	213	204	220
455	321	318	296	349	236	211	204	218

the body growth of the animals as inbreeding advanced. The females of the twenty-fifth generation (graph D) were, on the whole, slightly heavier than were the females of the preceding generation group (graph C).

Table 6 gives data showing the average body weights at different age periods of males and of females belonging to various generation groups of the B series of inbreds.

The data given in table 6 served as the basis of construction for the graphs shown in figure 3 and in figure 4.

A comparison of the graphs in figures 3 and 4 with the corresponding graphs in figures 1 and 2 shows that there was very little difference between the two inbred series (A and B) as regards the body-weight increase with age in the animals of the various generation groups. In the B series, as in the A series, males and

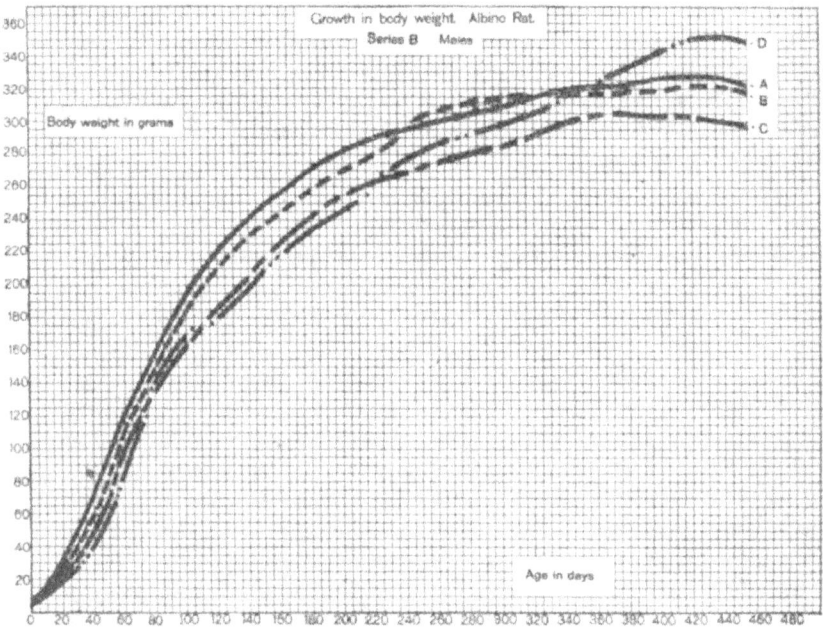

Fig. 3 Graphs showing the increase in the weight of the body with age for males belonging to various generation groups of the B series of inbred rats (data in table 6; lettering as in fig. 1).

females in the sixteenth to the eighteenth generation groups (graph A) were heavier animals at any given age than were those of subsequent generations; while the rats of the twenty-second to the twenty-fourth generation groups showed a much less vigorous growth than did the animals in the earlier groups. The rats in the twenty-fifth generation of the B series increased in body weight very slowly during the adolescent period, as the position of graph D in figures 3 and 4 indicates; but in the adult state their growth was much more vigorous, and their body weights, especially those of the males, compare favorably with the weights of the animals in the group comprising the rats of the sixteenth to the eighteenth generations (graph A).

An examination of figures 1 to 4 brings out one fact of considerable interest: all of the graphs have the same general form, although they vary somewhat in height. As the form of these graphs is practically the same as that of the growth graphs for

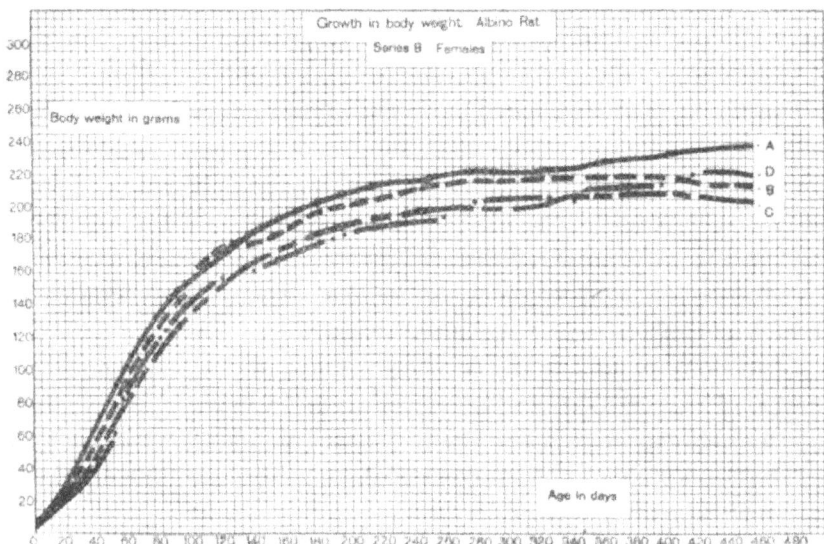

Fig. 4 Graphs showing the increase in the weight of the body with age for females belonging to various generation groups of the B series of inbred rats (data in table 6; lettering as in fig. 1).

stock Albinos as determined by Donaldson ('06) and others, it follows that close inbreeding, continued through many generations, does not alter the character of the growth graph for the albino rat. Theoretically, it might be expected, perhaps, that long-continued inbreeding would cause a slowing up of the growth processes, since the animals totally lack the stimulus to growth that a condition of heterozygosis seems to give in many cases (East and Hayes, '12; Jones, '18). The body weights of the animals in the sixteenth to the twenty-fifth generations of the inbred strain tended to lag somewhat during early postnatal life (figures 7 and 8, graph B), but this was undoubtedly due to the action of environmental and nutritive conditions, not to inbreeding. Any agency influencing growth, whether it be beneficial or detrimental, naturally produces its greatest effect during the period when growth is normally most rapid and vigorous. Since unfavorable conditions of environment and of nutrition unquestionably limited the extent of body growth in the animals of the later generations of the inbred strain, it is very probable that these factors also lessened the rate of growth during the early life of the individuals. If body growth in the inbred rats of future generations is retarded during the adolescent period, although the environmental and nutritive conditions under which the animals live are such that they produce rapid and vigorous growth in outbred stock Albinos, the change in the rate of growth can be ascribed to the effects of inbreeding. As far as the experiment has gone at present, the evidence does not warrant the conclusion that inbreeding per se has altered the form of the growth graph to any appreciable extent.

The body-weight data for the animals in various generation groups of the two inbred series, as given in table 5 and in table 6, were combined in order to show the weight increase with age in the individuals of the inbred strain as a whole. The combined data are shown in table 7.

The data in table 7 are not presented graphically, since there was such a close agreement between the corresponding records for the various generation groups of the two series that graphs constructed from the combined data would not differ materially from those given for the separate series (figs. 1 to 4).

Table 8 gives data showing the increase in the weight of the body with age for all of the individuals in the sixteenth to the twenty-fifth generations of the A series of inbreds for which growth records were taken; table 9 shows similar data for individuals of the B series.

A comparison of the data in table 8 with corresponding data in table 9 shows that the rats in the two inbred series were much alike as regards the rate and extent of their growth in body weight. To show this similarity more clearly, weight data for the males of the two series are presented graphically in figure 5.

TABLE 7

*Showing the average body weights at different ages of inbred rats of the two
series (A, B) separated into groups according to the generation
to which the individuals belonged*

AGE	MALES				FEMALES			
	Genera-tions 16–18	Genera-tions 19–21	Genera-tions 22–24	Genera-tion 25	Genera-tions 16–18	Genera-tions 19–21	Genera-tions 22–24	Genera-tion 25
days								
13	19	19	18	18	18	18	17	17
30	47	43	44	43	45	41	42	41
60	124	112	98	107	103	97	88	94
90	183	165	142	157	148	140	125	129
120	225	211	179	193	173	171	146	152
151	253	240	210	225	191	186	169	168
182	274	266	235	244	205	199	184	178
212	289	283	252	268	212	205	192	188
243	298	299	265	279	217	214	196	192
273	313	307	279	291	223	217	198	198
304	318	308	285	301	222	218	200	206
334	326	314	294	314	224	219	205	208
365	328	318	300	324	229	221	208	215
395	312	318	300	332	231	218	207	222
425	336	317	297	339	232	215	205	229
455	329	317	294	332	234	214	203	223

TABLE 8

*Showing the increase in the weight of the body with age for 179 males and for
130 females belonging in the sixteenth to the twenty-fifth
generations of the A series of inbred rats*

AGE	MALES				FEMALES			
	Body weight			Number of in-dividuals	Body weight			Number of in-dividuals
	Average	Highest	Lowest		Average	Highest	Lowest	
days	*grams*	*grams*	*grams*		*grams*	*grams*	*grams*	
13	18	24	14	179	17	24	14	130
30	43	54	36	179	42	57	33	130
60	109	205	58	179	95	158	62	130
90	161	268	92	179	135	187	80	121
120	203	294	128	179	162	218	108	125
151	232	321	163	178	182	235	133	123
182	256	361	196	178	196	238	162	122
212	274	382	192	170	203	246	159	116
243	285	404	199	162	208	268	157	110
273	298	432	215	148	212	268	178	100
304	302	413	213	141	215	279	169	95
334	308	410	213	127	216	273	174	91
365	314	418	223	116	220	283	168	86
395	308	421	231	103	220	305	164	80
425	313	485	227	85	218	298	169	68
455	311	447	223	75	214	269	162	58

The relative position of the graphs in figure 5 shows that during the early growth stages males of the B series of inbreds were slightly heavier at any given age than were the males of the A series; in the period from 100 to 300 days the advantage in body weight was with the males of the A series; beyond this age males of the B series were again the heavier. In the adult state the space between the graphs represents a difference of only about

TABLE 9

Showing the increase in the weight of the body with age for 117 males and for 180 females belonging in the sixteenth to the twenty-fifth generations of the B series of inbred rats

AGE	MALES				FEMALES			
	BODY WEIGHT			Number of in-dividuals	BODY WEIGHT			Number of in-dividuals
	Average	Highest	Lowest		Average	Highest	Lowest	
days	*grams*	*grams*	*grams*		*grams*	*grams*	*grams*	
13	19	24	15	117	18	22	14	180
30	46	62	36	117	44	60	34	180
60	111	147	64	117	96	137	63	180
90	163	230	110	117	136	188	98	161
120	201	281	153	117	162	218	122	169
151	230	326	165	117	179	236	136	164
182	255	358	189	116	186	247	143	176
212	270	367	195	115	199	250	157	163
243	285	392	219	113	205	261	169	163
273	294	415	227	106	208	277	168	148
304	300	410	236	104	209	290	172	148
334	311	459	258	96	213	287	181	136
365	315	460	259	85	216	280	180	126
395	317	449	239	78	216	293	177	114
425	319	455	246	64	215	293	171	99
455	315	450	238	56	213	279	168	78

2 per cent in the average body weights of the two groups of animals.

Graphs showing the increase in the weight of the body with age for females of the two inbred series are shown in figure 6. These graphs are based on data given in table 8 and in table 9.

In figure 6, as in figure 5, the graphs lie very close together throughout their entire course. Females in the B series of inbreds were slightly heavier animals than those in the A series

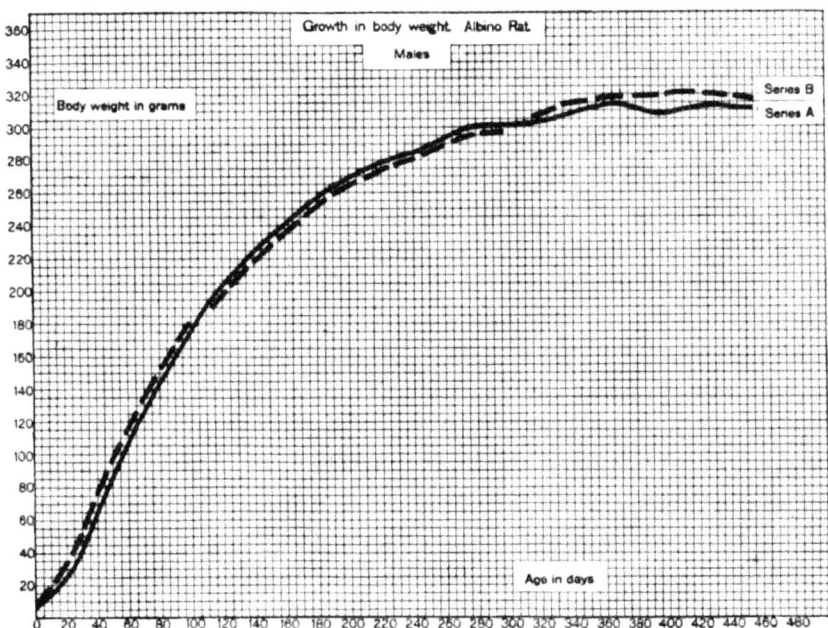

Fig. 5 Graphs showing the increase in the weight of the body with age for males belonging in the sixteenth to the twenty-fifth generations of the two series (A and B) of inbred rats (data in table 8 and in table 9).

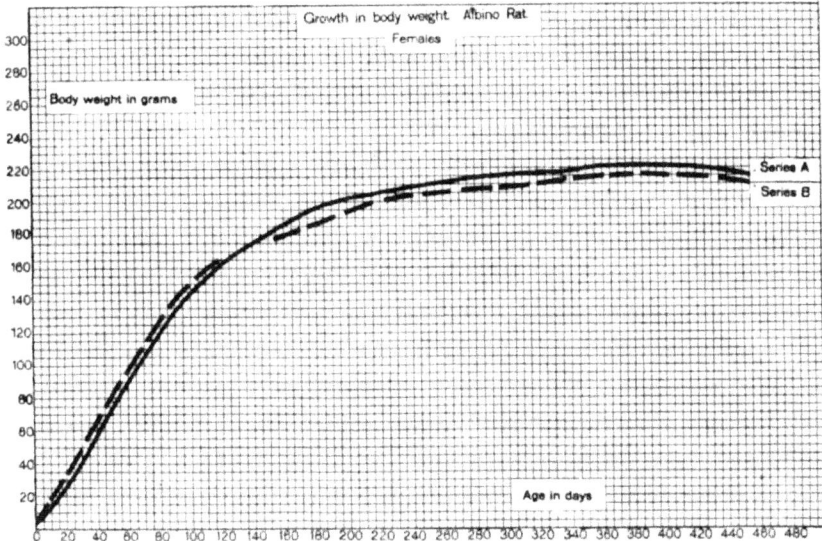

Fig. 6 Graphs showing the increase in the weight of the body with age for females belonging in the sixteenth to the twenty-fifth generations of the two series (A and B) of inbred rats (data in table 8 and in table 9).

149

during early life, but in the adult state this relation was reversed
and the females in the A series were about 2 per cent heavier,
as the graphs in figure 6 indicate.

In the seventh to the fifteenth generations of the inbred strain,
also, the animals of the two series had about the same average
body weight at corresponding age periods, although, as a group,
the individuals of the B series were slightly heavier (King, '18;
tables 11 and 12). Throughout the period of over nine years
that this experiment has been in progress, therefore, body growth
in the individuals of the one inbred series has closely paralleled
that of the individuals in the other series. If the varying conditions of environment and of nutrition to which the animals
of the inbred strain have been subjected have had any influence
on the heritable factors on which growth depends, it is evident
that they have acted on the animals of both series in a similar
way. I am strongly inclined to the opinion that environmental
and nutritive conditions do not influence genetic growth factors
directly, but that they act by either stimulating or retarding
the growth processes.

Body-weight data for a total of 606 individuals, 296 males and
310 females, belonging in the sixteenth to the twenty-fifth generations of the inbred strain are given in table 10. Reference to
this table, which is a combination of the data in table 8 and in
table 9, will be made later.

In connection with another problem I have recently taken a
series of body-weight records for a second group of outbred stock
Albinos. Supposedly these rats represented the best stock in
our colony at the time that the investigation was begun (1916),
as care was taken to select for breeding the largest and apparently
the most vigorous individuals from the large number available
for this purpose. These stock Albinos were reared simultaneously with, and under the same environmental and nutritive
conditions, as the inbred rats of the twenty-first to the twenty-
fifth generations. The body-weight data for these animals are
given in table 11.

A comparison of the body-weight data for the stock Albinos
(table 11) with that for the inbred group (table 10) shows that

the inbred rats, both males and females, were much heavier than the stock rats at every age for which records were taken. Not only were the animals in this stock series very inferior in size to those in the first stock series reared in 1913 to 1915 as controls for the inbred strain (King, '15; table 3), but their average body weights during adult life were no greater than those of the rats in the first six generations of the inbred strain which suffered severely from malnutrition (King, '18; table 3).

TABLE 10

Showing the increase in the weight of the body with age for 296 males and for 310 females belonging in the sixteenth to the twenty-fifth generations of the inbred rats (Series A and B combined)

AGE	MALES				FEMALES			
	BODY WEIGHT			Number of individuals	BODY WEIGHT			Number of individuals
	Average	Highest	Lowest		Average	Highest	Lowest	
days	*grams*	*grams*	*grams*		*grams*	*grams*	*grams*	
13	18	24	14	296	18	24	14	310
30	44	62	36	296	43	60	33	310
60	110	205	58	296	95	158	62	310
90	161	268	92	296	136	188	80	282
120	202	294	128	296	162	218	108	294
151	232	326	163	295	180	236	133	287
182	255	361	189	294	187	247	143	298
212	272	382	192	285	201	250	157	279
243	285	404	199	275	206	268	157	273
273	296	432	215	254	210	277	168	248
304	301	413	213	245	211	290	169	240
334	310	459	213	223	214	287	174	227
365	314	460	223	201	218	283	168	212
395	312	449	231	181	218	305	164	194
425	315	485	227	149	216	298	169	167
455	312	450	223	131	213	293	162	136

To facilitate a comparison between the body growth of inbred rats belonging in various generation groups and that of outbred stock Albinos, graphs showing the weight increase with age in two groups of inbred rats and in two groups of stock rats are given in figure 7 and in figure 8.

Growth graphs for various series of male rats are shown in figure 7.

In figure 7, graph A runs considerably above all of the other graphs, except at the thirteen-day period, thus showing that the growth of the males in the seventh to the fifteenth generations of the inbred strain was exceptionally vigorous. Males in the sixteenth to the twenty-fifth generations were relatively small: in the adult state their average body weights were about 9 per

TABLE 11

Showing the increase in the weight of the body with age and the coefficients of variability for 165 males and for 139 females belonging to a series of stock albino rats that were reared under the same environmental and nutritive conditions as the inbred rats belonging in the twenty-first to the twenty-fifth generations

	MALES			FEMALES		
AGE	Average body weight	Coefficients of variability	Number of individuals	Average body weight	Coefficients of variability	Number of individuals
days	grams			grams		
13	15	15.8±0.92	165	17	16.0±0.84	139
30	40	18.4±1.01	165	39	17.6±1.04	139
60	94	21.3±0.83	150	83	20.2±0.83	131
90	126	20.0±0.97	149	116	17.4±0.75	122
120	173	19.6±0.76	149	137	14.9±0.66	118
151	195	18.1±0.71	149	152	12.0±0.52	120
182	213	15.9±0.63	147	164	11.3±0.51	111
212	226	18.0±0.71	143	171	13.7±0.65	102
243	232	17.6±0.61	137	174	13.1±0.61	105
273	239	18.3±0.76	129	185	13.0±0.74	101
304	243	19.6±0.87	116	186	14.4±0.71	94
334	247	17.9±0.89	108	189	14.3±0.72	87
365	254	15.8±0.77	94	188	14.4±0.74	86
395	258	15.6±0.76	75	195	15.4±0.86	71
425	263	19.1±1.24	54	192	14.8±0.93	57
455	269	18.0±1.35	39	195	15.2±1.02	49
		18.0±0.85			14.8±0.76	

cent less than those of the males in the earlier generations, as the position of graph B indicates.

A comparison of graph B with graph C in figure 7 shows that the body-weight increase with age in the males of the later generations of the inbred strain was, on the whole, very similar to that in the males of the series of stock Albinos reared in 1913 to 1915 as controls for the inbred strain: stock males grew somewhat more vigorously during the adolescent period, but they were not as heavy as the inbred males in adult life. Since the inbred males were fully as large as the males in the stock series that had been reared under much more favorable conditions of environment and of nutrition, it is evident that continued inbreeding

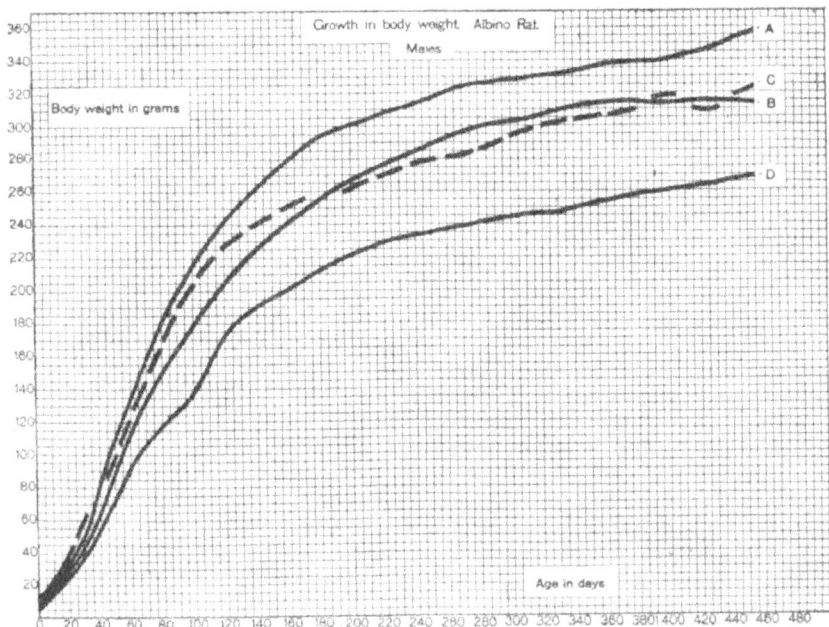

Fig. 7 Graphs showing the increase in the weight of the body with age for males belonging to four series. A, graph for males of the seventh to the fifteenth generations of the inbred strain (series A, B); B, graph for males of the sixteenth to the twenty-fifth generations of the inbred strain (series A, B); C, graph for males of the selected series of stock Albinos reared in 1913 to 1915 as controls for the inbred strain; D, graph for males of the stock series reared simultaneously with the individuals of the twenty-first to the twenty-fifth generations of the inbred strain (data in table 10 and in table 11 of the present paper and in table 13 of 'Studies on inbreeding I;' King, '18).

has not produced a deterioration in the original stock as regards the normal weight increase with age. The males in the seventh to the fifteenth generations of the inbred strain were much superior in body weight to outbred stock males reared under similar environmental and nutritive conditions (compare graph A with graph C in figure 7). Likewise, inbred males of the sixteenth to the twenty-fifth generations, living for the most part under the handicap of inadequate nutrition, were considerably heavier at all ages than the males in a stock series that were reared simultaneously with them, as a comparison of graph B with graph D in figure 7 shows. The space between these graphs, at the 200-day period, indicates a difference of about 17 per cent in favor of the males of the inbred group.

Growth graphs for various groups of female rats are shown in figure 8.

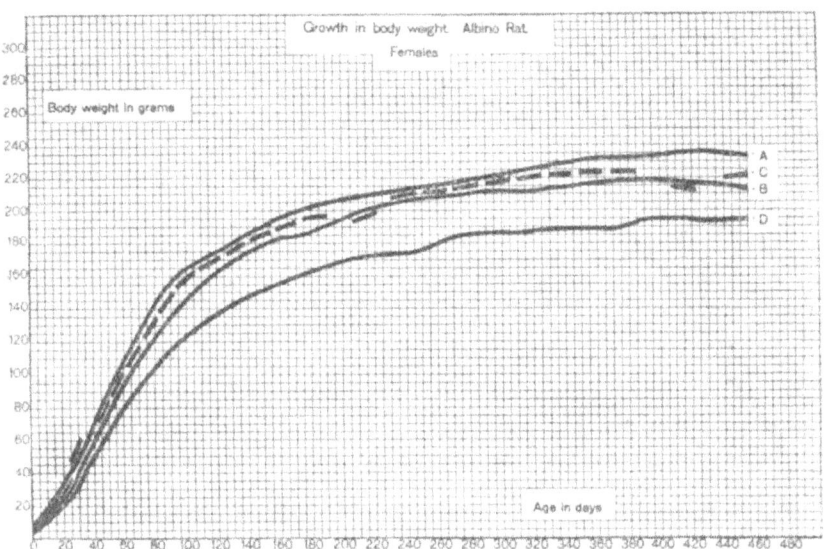

Fig. 8 Graphs showing the increase in the weight of the body with age for females belonging to four series (data and lettering as in table 7).

The growth graphs for various groups of females, shown in figure 8, have the same relative positions as have the graphs for the corresponding groups of males (fig. 7), but they lie somewhat closer together. Inbred females of the seventh to the fifteenth generations, as graph A shows, were heavier at all ages (except thirteen days) than the females of the other groups; in the adult state their average body weights were about 2 per cent greater than those of the inbred females belonging in subsequent generations (graph B). Body weight increase with age in females of the sixteenth to the twenty-fifth generations of the inbred strain closely followed that of the females in the first series of stock controls (compare graph B with graph C in figure 8). The animals in both of these latter groups were about 14 per cent heavier in adult life than the females in the stock series reared during the past two years (graph D).

In explanation of the remarkably vigorous growth of the animals in the seventh to the ninth generations of the inbred strain it was suggested in the first paper of this series (King, '18) that: "favorable nutritive conditions following a period of semi-starvation greatly increased metabolic activity and so stimulated the growth impulse that the animals attained an unusually large size. After the maximum effect of the stimulus had passed there was a gradual decline to more normal conditions of metabolism and a corresponding decrease in the average size of the individuals." Rats seem to be particularly sensitive to changes in food conditions, more so than is generally supposed, and only by feeding them constantly on a proper diet can their normal weight and fertility be maintained. In light of the valuable researches of McCollum ('18) and his associates, it is evident that the 'scrap' food that the rats received during the period when they exhibited their maximum growth and fertility not only furnished a well-balanced ration as regards the basic food stuffs, but that it also gave a sufficient quantity of the essential accessory foods, 'fat-soluble A' and 'water-soluble B,' to greatly stimulate the growth processes. The experimental diets recently used in our colony have very evidently been deficient in 'fat-soluble A.' As a result the rats have shown marked evidence of malnutrition, although they have received an abundance of food. By rectifying the mistakes of the past and feeding the animals on a properly balanced ration, it is hoped that body growth will again respond to the stimulus of adequate nutrition and that it will be possible to obtain inbred animals that are as large as those in the seventh generation. As after twenty-five generations of brother and sister matings the animals in the inbred strain were fully as large as were the best stock animals obtainable, it is evident that close inbreeding does not inevitably

cause a decrease in body size, as Darwin ('75, '78), Crampe ('83),
Ritzema-Bos ('93, '94), and others have asserted. Inadequate
nutrition, seemingly, is far more detrimental to body growth than
is close inbreeding, even when continued over many generations.

VARIABILITY IN THE BODY WEIGHTS OF INBRED RATS

At the end of fifteen generations of brother and sister matings
the rats in the inbred strain were over 96 per cent homozygous,
according to the calculations of Fish ('14). Animals of the later
generations, which had attained a degree of homozygosity probably greater than that ever before reached by any group of laboratory mammals, might be expected, perhaps, to show a very
great uniformity in body weight at different age periods, if the
body weight increase with age in the rat is entirely dependent
on the action of genetic growth factors. But just as the rate
and extent of body growth in this animal seems to be largely a
matter of environment and of nutrition, so also the variations
in body weights at different age periods are apparently greatly
influenced by these conditions. As it is impossible, at present,
to distinguish the variability due to environmental and nutritive
action from that resulting from a difference in the genetic factors
for body growth, one can only calculate the total amount of
variability in given groups of animals and then, by comparison,
determine the relative variability of the groups. No very definite conclusions can be drawn regarding the effects of close
inbreeding on the variability in the body weight of the rat until
the animals can be kept under environmental and nutritive
conditions that are so uniform that their effect is practically
constant and therefore negligible.

In order to obtain some idea regarding the relative extent of
variability in the body weights of the animals in various generations of the inbred strain, coefficients of variability, with their
probable error, were determined for the body weights of the
individuals in the sixteenth to the twenty-fifth generations of
each of the two inbred series and for the weights of the animals
in the two series combined (A, B). These coefficients, with

their probable error, were calculated from the data summarized in tables 8, 9, and 10 according to the formulae given by Davenport ('14); they are shown in table 12.

During early postnatal life, as the coefficients in table 12 show, the females in both inbred series were slightly more variable in body weight than were the males, but after thirty days of age the males, as a rule, were the more variable. Variability

TABLE 12

Showing the coefficients of variation, with their probable error, for the body weights at different ages of the two series of inbred rats (sixteenth to the twenty-fifth generations, inclusive)

AGE	SERIES A		SERIES B		COMBINED SERIES (A, B)	
	Males	Females	Males	Females	Males	Females
days						
13	12.6±0.45	13.5±0.56	12.2±0.54	12.2±0.44	12.4±0.36	12.9±0.33
30	12.1±0.43	11.9±0.50	14.3±0.63	14.4±0.52	13.5±0.37	13.3±0.36
60	22.9±0.82	18.0±0.75	16.3±0.71	16.4±0.60	20.6±0.57	17.1±0.46
90	20.1±0.72	15.8±0.69	14.8±0.65	14.0±0.53	18.4±0.51	15.5±0.44
120	19.1±0.68	13.7±0.58	14.3±0.63	11.9±0.44	16.1±0.45	12.7±0.35
151	14.5±0.52	10.2±0.45	13.7±0.60	10.7±0.39	14.2±0.39	10.5±0.30
182	13.3±0.48	9.1±0.39	12.6±0.59	12.3±0.44	13.1±0.36	10.4±0.29
212	13.3±0.49	9.6±0.43	12.0±0.53	8.7±0.33	12.8±0.43	9.2±0.39
243	13.3±0.50	9.8±0.45	11.3±0.51	8.5±0.32	12.4±0.36	9.3±0.27
273	12.6±0.49	9.9±0.47	10.9±0.50	9.1±0.35	11.8±0.35	9.4±0.29
304	11.3±0.45	10.8±0.53	10.2±0.48	8.6±0.34	11.2±0.34	9.7±0.30
334	12.3±0.52	10.1±0.50	11.2±0.55	7.9±0.32	11.7±0.37	8.9±0.28
365	12.3±0.54	10.3±0.53	12.0±0.62	9.0±0.38	12.2±0.45	9.6±0.31
395	12.9±0.61	11.3±0.62	12.4±0.67	9.8±0.44	12.7±0.45	10.5±0.36
425	14.2±0.73	10.2±0.59	12.9±0.77	10.5±0.50	13.8±0.54	10.7±0.39
455	14.3±0.79	11.7±0.73	13.6±0.69	11.5±0.62	14.1±0.59	11.6±0.47
Average........	14.4±0.58	11.6±0.55	12.8±0.60	10.9±0.43	13.8±0.43	11.3±0.35

was at its maximum for both sexes at the sixty-day period, and then tended to decrease with advancing age for some time. In table 12 the average coefficient for the male group in each of the two inbred series, taking all ages together, exceeds that for the corresponding group of females by over two points. Since this difference is over three times the probable error, it is sufficiently large to indicate that the males had a greater range of vari-

ability in body weight than had the females. Coefficients of
variability for the body weights of the individuals in the earlier
generations of the inbred strain (King, '18; table 15), and also
those for various series of stock Albinos (Jackson, '13; King, '15),
all show that the males are more variable than the females. Such
a relation between the sexes as regards the variability in their
body weights would seem to be a characteristic of the albino
strain of rats in general, and from the results obtained in the
present study it is evident that this relation has not been changed
by twenty-five generations of close inbreeding.

Males in the sixteenth to the twenty-fifth generations of the
A series of inbreds had a somewhat greater range of variability
in body weight than had the males of the B series, judging from
the relative size of the coefficients for the two series as given in
table 12. Between the average coefficients for the two series
there is a difference of 1.6 points in favor of the males of the
A series; a similar relation between the two series existed also at
an earlier period (King, '18; table 15). Throughout all generations of the inbred strain, therefore, the range of variability
in body weights was greater in the males of the A series than in
those of the B series. This difference persisted even during the
periods when body growth and variability were greatly influenced by environmental and nutritive conditions.

A comparison between corresponding coefficients for the females of the two inbred series (table 12) shows that, as a rule,
the females of the A series were slightly more variable in body
weight at different age periods than were the females of the B
series, but, taken as a whole, the one group of females was about
as variable as the other, since the difference between the average
coefficients for the two groups is only 0.7 point. As the study
of variability in the females of the earlier generations of the
inbred strain led to the conclusion that "the range of variability
in body weights was practically the same for the females of the
two inbred series," it is evident that long-continued inbreeding
has not altered the relative variability of the females in the two
inbred series any more than it has that of the males.

Table 12 shows that in each inbred series the coefficients of variability for both sexes decrease in size with advancing age until the animals attained an age of about 300 days, and then tend to become somewhat larger; a similar change in the size of the coefficients at various age periods was also noted for the animals in the earlier generations of the inbred strain as well as for those in the two stock series reared as controls. After reaching the height of their reproductive activity at the age of from seven to ten months, certain individuals, especially males, tend to accumulate an excess of adipose tissue; while other individuals, even members of the same litter, will show little change in body weight for a period of several months, or they may even decline steadily in body weight although they are apparently in good physical condition. The increased variability in the body weights of older rats is, therefore, due in great part to the accumulation of a greater or less amount of adipose tissue; it is not a growth phenomenon comparable to that shown during early postnatal life.

In order to make a closer analysis of the relative variability in the body weights of animals in successive generations of the inbred strain, coefficients of variability were calculated from the body-weight data for the animals in three generations combined as summarized in table 7. This series of coefficients is shown in table 13.

In table 13 the average coefficients for the male groups comprising the individuals of the sixteenth to the twenty-fourth generations vary by less than one point, so it is evident that in the later generations of the inbred strain the variability in the body weights of the males did not decrease with the advance of inbreeding, as was the case in the earlier generations (King, '18; table 16). The series of coefficients for the males of the twenty-fifth generation are, as a rule, smaller than the corresponding coefficients for the males of the preceding generation group. But the difference between the average coefficients for the two groups is less than three times the probable error, so it cannot be considered as significant, especially as the number of body-weight records used in calculating the coefficients for the animals

TABLE 13

Showing the coefficients of variation, with their probable error, for the body weights at different ages of inbred rats of the two series (A, B) separated into groups according to the generation to which the individuals belonged

AGE	MALES				FEMALES			
	Generations 16-18	Generations 19-21	Generations 22-24	Generation 25	Generations 16-18	Generations 19-21	Generations 22-24	Generation 25
days								
13	12.5±0.66	13.7±0.72	12.8±0.61	7.7±0.64	13.5±0.72	13.8±0.67	12.8±0.62	7.1±0.58
30	13.3±0.70	13.1±0.69	13.4±0.64	6.9±0.57	14.7±0.79	11.8±0.57	13.6±0.66	7.6±0.62
60	17.9±0.94	17.0±0.90	15.1±0.72	18.8±0.76	15.8±0.80	13.7±0.67	18.3±0.89	14.6±1.20
90	14.2±0.75	15.6±0.82	17.5±0.84	12.1±1.00	11.1±0.64	12.3±0.64	15.7±0.78	11.1±0.94
120	11.8±0.63	13.9±0.73	14.3±0.68	10.5±0.87	11.1±0.59	10.3±0.52	11.6±0.59	8.2±0.67
151	12.4±0.65	11.9±0.63	12.0±0.57	10.2±0.85	10.2±0.54	7.9±0.39	9.0±0.46	8.8±0.77
182	11.4±0.61	11.1±0.59	12.1±0.58	9.6±0.80	9.4±0.51	6.4±0.33	8.6±0.43	6.7±0.56
212	11.1±0.61	10.7±0.58	13.2±0.64	10.1±0.83	8.4±0.47	6.8±0.35	8.6±0.43	7.6±0.66
243	10.4±0.59	10.9±0.60	12.2±0.59	9.9±0.82	8.4±0.49	7.0±0.38	8.0±0.40	7.1±0.62
273	10.3±0.63	11.3±0.64	11.4±0.56	9.9±0.85	9.8±0.59	6.9±0.38	7.1±0.38	5.9±0.52
304	9.9±0.61	10.1±0.59	10.3±0.52	11.0±1.00	9.6±0.60	8.1±0.55	8.1±0.42	7.8±0.71
334	10.4±0.69	11.8±0.71	10.6±0.56	10.9±0.99	9.3±0.59	7.6±0.44	8.1±0.44	6.3±0.58
365	11.4±0.85	12.6±0.81	11.3±0.61	10.6±0.98	9.7±0.66	8.8±0.53	8.6±0.48	7.2±0.66
395	15.9±1.37	12.4±0.85	10.4±0.58	10.8±0.97	10.4±0.76	8.9±0.56	8.8±0.52	11.1±1.00
425	11.2±1.05	15.8±1.22	10.6±0.65	12.2±1.21	9.8±0.80	8.3±0.58	9.6±0.56	9.9±1.05
455	13.3±1.35	14.1±1.12	11.7±0.78	13.0±1.32	11.9±1.25	9.4±0.72	10.7±0.67	8.6±0.99
Average.........	12.3±0.79	12.8±0.76	12.4±0.63	10.9±0.86	10.8±0.67	9.2±0.51	10.4±0.55	8.5±0.76

of a single generation was only about one-third of that used for a group of three generations.

The average coefficients for the three groups of females comprising the animals in the sixteenth to the twenty-fourth generations of the inbred strain are all lower than those for the corresponding groups of males (table 13), and they also fail to show a significant decrease in size as inbreeding advanced. The average coefficient for the body weights of the females in the twenty-fifth generation is considerably smaller than that for any of the three generation groups, but here also no definite conclusion seems warranted, since the small number of records on which the coefficients are based may be responsible in great measure for the result.

The animals in the seventh to the fifteenth generations of the inbred strain lived under environmental and nutritive conditions that were fairly uniform and seemingly very favorable to growth and to fertility. The body weights of these individuals showed a slow decrease in variability with the advance of inbreeding, as the relative size of their coefficients of variability indicates (King, '18; table 16). During early life the rats in the sixteenth and seventeenth generations lived under the same environmental and nutritive conditions as the animals of the preceding generations, and at this time they were all seemingly somewhat less variable in body weight than were the individuals in the fifteenth generation. Before the weight records for these rats were completed, a change in diet became necessary, as 'scrap' food of the required quality and quantity could no longer be obtained. The effects of the change in food became very apparent in the course of a few weeks, and, as individual rats responded differently to the altered conditions of nutrition, there was a marked increase in the variability of the body weights in the animals of all ages. When the coefficients of variability were calculated from the series of body-weight data obtained for the animals in the sixteenth to the eighteenth generations, they were found to be somewhat larger than those for the animals in the fifteenth generation, as was expected from the observed appearance of the animals. The animals in the later generations of the inbred strain have shown a variability in body weights considerably greater than that found in any group of inbred animals since the tenth generation.

By comparing the corresponding coefficents for the two series of outbred stock Albinos that were reared in the colony on different diets, one can determine whether the variability in the body weights of these animals was influenced by the nutritive conditions under which they lived. By a further comparison of these coefficients with those for the animals in the later generations of the inbred strain, it will be possible to determine whether the increase in the variability of the inbred animals was due to altered conditions of nutrition or to the effects of long-continued inbreeding.

All of the stock Albinos reared in 1913 to 1915 as controls for the inbred series were fed on 'scrap' food. As has already been recorded (King, '15; table 4), the coefficients of variability for the body weights of the fifty males in this series range from 10.2 to 17.0, with an average of 13.6 for the entire group, taking all ages together; coefficients for the fifty females vary from 8.9 to 15.7, with an average of 11.5 for the entire group.

The second series of stock controls was reared in 1916 to 1918 simultaneously with the inbred rats of the twenty-first to the twenty-fifth generations, and they, as the inbred rats, were fed on various experimental diets. These stock Albinos came from the same general stock colony that furnished animals for the first series of controls, so the coefficients for the two series are strictly comparable. An examination of the coefficients for the body weights of the rats in this control series, as given in table 11 of the present paper, shows that all of them are much larger than the corresponding coefficients for the animals of the first stock series, while the difference between the average coefficients for the two series is over four times the probable error. It is evident, therefore, that the rats in the second series of stock controls were much more variable in their body weights at all age periods than were the animals in the first stock series. Since both of these stock series were outbred, the increased variability in the animals of the second series cannot be attributed to the effects of inbreeding; nor can it be ascribed to a difference in the genetic constitution of the two series of animals, since no new 'blood' was introduced into the general stock colony from 1913 to 1917. From the evidence given, one seems warranted in assuming that the marked difference in the variability of the two series of stock animals was due, in great part, to the effects of changed conditions of nutrition which so greatly influenced the body growth of the individuals in the second series. It is probable also that the extremes of temperature to which many of these rats were subjected also affected their variability in body weight to some extent, although the effects of temperature changes were very much less than those of nutrition.

Since the variability in the body weights of outbred stock Albinos was seemingly greatly affected by nutritive and environmental factors, one would naturally conclude that these factors would likewise influence the variability in the body weights of inbred animals reared simultaneously with and under the same conditions as the stock Albinos. The increased variability in the inbred animals of the sixteenth to the twenty-fifth generations is, on this assumption, the result of environmental and nutritive action, and it cannot be cited in support of Walton's ('15) contention that continued inbreeding tends to increase variability. It is interesting to note in this connection that a comparison between the average coefficients for various groups of inbred rats and those for stock Albinos indicates that changed conditions of nutrition produced a much greater effect on the variability in the body weights of stock Albinos than it did on that of the animals in the later generations of the inbred strain.

In this experiment, owing to the action of environment and of nutrition, it is impossible to determine the changes, if any, that inbreeding per se produced on the variability in the body weights of the animals in the later generations of the inbred strain. This study of variability is of value, therefore, mainly because it shows that in the later generations of inbreds there existed between the two series (A and B), and between the two sexes, the same relative variability in body weights as that found in the earlier generations. Twenty-five generations of brother

and sister matings have not, seemingly, altered the relative
variability in the strain, whether the total amount of variability
has been influenced by inbreeding cannot be determined until
it is possible to rear a number of generations of these animals
under uniform conditions of environment and of nutrition.

GENERAL CONCLUSIONS

As a whole, this experiment has shown that the closest form
of inbreeding possible in mammals, the mating of brother and
sister from the same litter, is not necessarily inimical either to
body growth, to fertility, or to constitutional vigor, provided
that only the best animals from a relatively large number are
used for breeding purposes. Selection, seemingly, is able to
hold in check any tendency that inbreeding may have to bring
out the undesirable, latent traits inherent in the strain.

In the course of this investigation it has been shown that
adverse conditions of environment and of nutrition produce far
more detrimental effects on growth and fertility in the albino
rat than does inbreeding. These factors, apparently, do not
alter the genetic constitution of the individual, since the animals
soon resume their normal growth and fertility when environmental and nutritive conditions are again favorable.

The sex ratio in the rat is seemingly a character that is amenable to selection, since through this process the inbred strain
has been separated into two lines: one line (A) showing a high
sex ratio, the other line (B) showing a low sex ratio. The effects
of selection on the sex ratio seem to be limited, however, since
there has been no cumulative effects of the selection, although
the two lines have been kept distinct for eighteen successive
generations. Whether it will be possible to change the sex
ratio in the two lines by reversing the selection is the chief problem in view in the continuation of this work.

Throughout the entire course of this investigation there has
been a great similarity between the two inbred series as regards
the variability in the body weights of the animals at different
age periods. In the earlier generations the variability in body

weights seemed to decrease with the advance of inbreeding, but in the later generations the variability was greatly influenced by environmental and nutritive conditions.   Until these latter factors can be controlled, it will not be possible to draw any definite conclusions regarding the effects of inbreeding per se on the variability in body weights.

SUMMARY

1. The data given in the present paper show the growth and variability in the body weights of 296 males and of 310 females belonging in the sixteenth to the twenty-fifth generations of two series (A and B) of albino rats that were inbred, brother and sister from the same litter.

2. Owing to economic conditions, many of these rats were not reared under very favorable conditions of environment and of nutrition, and in consequence they did not grow as rapidly nor did they attain as great a maximum body weight as did the individuals in the earlier generations of this inbred strain.

3. In every generation from the sixteenth to the twenty-fifth the males were heavier than the females at all age periods after thirty days (tables 1 to 4).   This result agrees with the finding for the inbred rats of the earlier generations, and also with that for various series of stock Albinos.   Apparently, therefore, long-continued inbreeding has not changed the normal body-weight relations of the sexes at any age period for which records have been taken.

4. In the A series of inbreds the rate and extent of growth in body weight were much the same as those in the B series of inbreds: in the adult animals there was a difference of only about 2 per cent in the average body weights of corresponding groups of males and females in the two series (tables 8 and 9; fig. 5 and 6).

5. Close inbreeding for twenty-five generations has not altered the form of the growth graph for the albino rat to any extent.

6. Rats belonging to the later generations of the inbred strain were not as heavy at any age period as were the animals in the earlier generations, but they were much superior in body weight to stock Albinos reared under similar conditions of environment and of nutrition (figs. 7 and 8).

7. Individuals in the sixteenth to the twenty-fifth generations of the inbred strain had about the same average body weight at different age periods as had the individuals of the stock controls reared in 1913 to 1915 under favorable conditions of environment and of nutrition (figs. 7 and 8; compare graph B with graph C). Seemingly, therefore, inbreeding has as yet produced no deterioration in the original Albino stock as regards the rate and extent of growth in body weight.

8. Variability in the body weights of the animals in the later generations of the inbred strain followed the same general trend as that in the animals of the earlier generations and in those of the two stock series studied: in both sexes it increased from birth to sixty days, and then decreased steadily until the animals were about 300 days of age, tending to rise again in older rats (table 12).

9. In the later generations of the inbred strain the males were more variable in body weight than the females. This result agrees with the finding for the animals of the earlier generations and for various series of stock Albinos.

10. In the inbred animals of the sixteenth to the twenty-fifth generations variability in body weights was relatively high, and it did not tend to decrease with the advance of inbreeding as in the earlier generation (table 13).

11. Outbred stock Albinos, reared simultaneously with and under the same environmental and nutritive conditions as the inbred rats of the twenty-first to the twenty-fifth generations, showed a variability in their body weights at all ages much greater than that in the animals of the earlier stock series reared under more favorable conditions of nutrition. It appears, therefore, that the increased variability in the body weights of the animals in the later generations of the inbred strain was due to the action of environment and of nutrition, not to the effect of continued inbreeding.

LITERATURE CITED IN STUDIES ON INBREEDING I TO IV

AHLFELD, DR. 1876 Über den Knabenüberschuss der älteren Erstgebärenden nebst einem Beitrage zum Hofacker-Sadler'schen Gesetze. Arch. Gynäkologie, Bd. 9.

ALLEN, EZRA 1917 Spermatogenesis in the albino rat. Abstract in Proc. Amer. Soc. Zoologists. Anat. Rec., vol. 11.
1918 Studies on cell division in the albino rat (Mus norvegicus var. alb.) III. Spermatogenesis: the origin of the first spermatocytes and the organization of the chromosomes, including the accessory. Jour. Morp., vol. 31.

ANDERSON, W. S. 1911 A study in heredity. Report sixth annual Ky. State Farmer's Institute.

B. 1906 Ein Beitrag zur Wirkung des Geschlechtsverkehrs zwischen Blutsverwandten. Kosmos, Bd. 3.

BALTZER, F. 1914 Die Bestimmung des Geschlechts nebst eine Analyse des Geschlechtsdimorphismus bei Bonellia. Mittheil. Zoöl. Sta. Neapoli, Bd. 22.

BANTA, A. M. 1916 Sex intergrades in a species of Crustacea. Proc. Nat. Acad. Sci., vol. 2.

BASSET, G. C. 1914 Habit formation in a strain of albino rats of less than normal brain weight. Behavior Monographs, vol. 2.

BIDDER, F. 1878 Über den Einfluss des Alters der Mutter auf das Geschlecht des Kindes. Zeitschr. Geburtshülfe u. Gynäkologie, Bd. 11.

BORN, G. 1881 Experimentelle Untersuchungen über die Entstehung der Geschlechtsunterschiede. Breslauer Ärzt. Zeitschr., Bd. 3.

BURCK, W. 1908 Darwin's Kreuzungsgesetz und die Grundlagen der Blütenbiologie. Rec. Trav. Bot. Néerl., Bd. 4.

CASTLE, W. E. 1896 The early embryology of Ciona intestinalis Flemming (L). Bull. Mus. Comp. Zool., vol. 27.
1903 The heredity of sex. Bull. Mus. Comp. Zool., vol. 40.
1916 Genetics and eugenics. Cambridge, Mass.
1916 a Size inheritance in guinea-pig crosses. Proc. Nat. Acad. Sci., vol. 2.
1916 b Variability under inbreeding and crossbreeding. Amer. Nat., vol. 50.

CASTLE, W. E., CARPENTER, F. W., et al. 1906 The effects of inbreeding, crossbreeding and selection upon the fertility and variability of Drosophila. Proc. Amer. Acad. Arts and Sci., vol. 41.

CASTLE, W. E., AND PHILLIPS, J. C. 1914 Piebald rats and selection. Carnegie Inst. Pub. no. 195, Washington, D. C.

CASTLE, W. E., AND WRIGHT, SEWALL 1916 Studies of inheritance in guinea-pigs and rats. I. An expedition to the home of the guinea-pig and some breeding experiments with material there obtained. Carnegie Inst. Pub. no. 241, Washington, D. C.

CHAPEAUROUGE, A DE 1909 Einiges über Inzucht und ihre Leistung auf verschiedenen Zuchtgebieten. Hamburg.

COLE, L. J., AND KIRKPATRICK, W. F. 1915 Sex ratios in pigeons, together with observations on the laying, incubation and hatching of the eggs. Bull. no. 162, Agri. Exper. Sta., R. I. State College.

CONROW, S. B. 1915 Taillessness in the rat. Anat. Rec., vol. 9.
1917 Further observations on taillessness in the rat. Anat. Rec., vol. 12.

COPEMAN, S. MONCKTON, AND PARSONS, F. G. 1904 Observations on the sex in mice. Preliminary paper. Proc. Royal Soc., vol. 73.

CRAMPE, H. 1883 Zucht-Versuche mit zahmen Wanderratten. I. Resultate der Zucht in Verwandtschaft. Landwirtschaftliche Jahrb., Bd. 12.
1884 Zucht-Versuche mit zahmen Wanderratten. II. Resultate der Kreuzung der zahmen Ratten mit wilden. Landwirthschaftliche Jahrb., Bd. 13.

CUÉNOT, L. 1899 Sur la determination du sexe chez les animaux. Bull. Sci. de la France et de la Belgique, T. 32.

DARWIN, CHARLES 1875 The variation of animals and plants under domestication. Second edition, London.
1878 The effects of cross and self fertilization in the vegetable kingdom. Second edition, London.

DAVENPORT, C. B. 1900 Review of von Guaita's experiments in breeding mice. Biol. Bull., vol. 2.
1914 Statistical methods with special reference to biological variation. John Wiley & Sons, New York.

DONALDSON, H. H. 1906 A comparison of the white rat with man in respect to the growth of the entire body. Boas Anniversary volume, New York.
1912 The growth of the brain. New York.
1915 The rat. Data and reference tables. Mem. Wistar Institute of Anat. and Biol., no. 6, Philadelphia.

DONCASTER, L. 1913 On an inherited tendency to produce purely female families in Abraxas grossulariata, and its relation to an abnormal chromosome number. Jour. Genetics, vol. 3.
1914 The determination of sex. Cambridge.
1914 a Chromosomes, heredity and sex. Quart. Jour. Micr. Sci., vol. 59.

DÜSING, KARL 1883 Die Factoren welche die Sexualität entscheiden. Inaug. Dissertation, Jena.
1884 Die Regulierung des Geschlechtsverhältnisses bei der Vermehrung der Menschen, Tiere und Pflanzen. Jen. Zeitschr. Naturwiss., Bd. 17.
1886 Die experimentelle Prüfung der Theorie von der Regulierung des Geschlechtsverhältnisses. Jen. Zeitschr. Naturwiss., Bd. 19.
1887-92 Die Regulierung des Geschlechtsverhältnisses bei Pferden. Landwirtschaftliche Jahrb., Bds. 16, 17, 21.

EAST, E. M., AND HAYES, H. K. 1911 Inheritance in maize. Bull. no. 167, Conn. Agri. Exper. Station.
1912 Heterozygosis in evolution and in plant breeding. Bull. no. 243. U. S. Dept. Agri.

EWART, J. C. 1910 Principles of breeding and the origin of domesticated breeds of animals]. Bureau Animal Industry, Washington, D. C.

FABRE-DOMENGUE, P. 1898 Unions consanguines chez les Columbins. L'intermédiare des Biol., T. 1.

FERRY, E. L. 1913 The rate of growth of the albino rat. Anat. Rec., vol. 7.

FISH, H. D. 1914 On the progressive increase of homozygosis in brother-sister matings. Amer. Nat., vol. 48.

GENTRY, N. W. 1905 · Inbreeding Berkshires. Proc. Amer. Breeders Assoc., vol. 1.

GOULD, H. N. 1917 Studies on sex in the hermaphrodite mollusk Crepidula plana. Jour. Exper. Zoöl., vol. 23.

GRIESHEIM, A. 1881 Über die Zahlenverhältnisse der Geschlechter bei Rana fusca. Arch. ges. Physiol., Bd. 26.

GUAITA, G. VON 1898 Versuche mit Kreuzungen von verscheidenen Rassen der Hausmaus. Ber. naturf. Gesellsch. zu Freiburg, Bd. 10.
1900 Zweite Mittheilung über Versuche mit Kreuzungen von verschiedenen Hausmausrassen. Ber. naturf. Gesellsch. zu Freiburg, Bd. 11.

GUYER, M. F. 1909 On the sex of hybrid birds. Biol. Bull., vol. 16.
1910 Accessory chromosomes in man. Biol. Bull., vol. 19.

HAMMOND, J. 1914 On some factors controlling fertility in domestic animals. Jour. Agri. Sci., vol. 6.

HATAI, S. 1907 Effect of partial starvation followed by a return to normal diet on the growth of the body and central nervous system of albino rats. Amer. Jour. Phys., vol. 17.

HAYES, H. K., AND JONES, D. F. 1917 The effects of cross- and self-fertilization in tomatoes. Report Conn. Agri. Exper. Sta.

HEAPE, W. 1899 Abortion, barrenness and fertility in sheep. Jour. Royal Agri. Soc., vol. 10.
1908 Note on the proportion of the sexes in dogs. Proc. Cambridge Phil. Soc., vol. 14.
1909 The proportion of the sexes produced by white and coloured peoples in Cuba. Phil. Trans. Royal Soc., vol. 200.

HERTWIG, R. 1906 Untersuchungen über das Sexualitätsproblem. Verhandl. deutsch. zool. Gesellsch.
1907 Weitere Untersuchungen über das Sexualitätsproblem. Verhandl. deutsch. zool. Gesellsch.

HIRSCH, M. 1913 Über das Verhältnis der Geschlechter. Centralbl. für Gynäkologie, Bd. 37.

HOSKINS, E. R. 1916 The growth of the body and organs of the albino rat as affected by feeding various ductless glands (thyroid, thymus, hypophysis and pineal). Jour. Exper. Zoöl., vol. 21.

HUTH, A. H. 1887 The marriage of near kin. 2nd Edition. London.

HYDE, R. R. 1914 Fertility and sterility in Drosophila ampelophila. I. Sterility in Drosophila with special reference to a defect in the female and its behavior in heredity. Jour. Exper. Zoöl., vol. 17.
1914 a Fertility and sterility in Drosophila ampelophila. II. Fertility in Drosophila and its behavior in heredity. Jour. Exper. Zoöl., vol. 17.

ISSAKOWITSCH, A. 1905 Geschlechtsbestimmenden Ursachen bei den Daphniden. Biol. Centralbl., Bd. 25.

JACKSON, C. M. 1912 On the recognition of sex through external characters in the young rat. Biol. Bull., vol. 23.
　　1913 Postnatal growth and variability of the body and of the various organs in the albino rat. Am. Jour. Anat., vol. 15.
　　1915 Changes in the relative weights of the various parts, systems and organs of young albino rats held at constant body weight by underfeeding for various periods. Jour. Exper. Zoöl., vol. 19.

JENNINGS, H. S. 1916 Heredity, variation and the results of selection in the uniparental reproduction of Difflugia corona. Genetics, vol. 1.

JOHANNSEN, W. 1909 Elemente der exakten Erblichkeitslehre. Jena.

JONES, D. F. 1918 The effects of inbreeding and crossbreeding upon development. Conn. Agri. Exper. Station, Bull. no. 207.

JORDAN, H. E. 1911 The spermatogenesis of the opossum. Arch. Zellforsch., Bd. 7.

KING, HELEN DEAN 1907 Food as a factor in the determination of sex in amphibians. Biol. Bull., vol. 13.
　　1911 Studies on sex determination in amphibians. IV. The effects of external factors, acting before or during the time of fertilization, on the sex ratio of Bufo lentiginosus. Biol. Bull., vol. 20.
　　1911 a The sex ratio in hybrid rats. Biol. Bull., vol. 21.
　　1912 Studies on sex determination in amphibians. V. The effects of changing the water content of the egg, at or before the time of fertilization, on the sex ratio of Bufo lentiginosus. Jour. Exper. Zoöl., vol. 12.
　　1915 Growth and variability in the body weight of the albino rat. Anat. Rec., vol. 9.
　　1915 a On the weight of the albino rat at birth and the factors that influence it. Anat. Rec., vol. 9.
　　1916 On the postnatal growth of the body and of the central nervous system in albino rats that are undersized at birth. Anat. Rec., vol. 11.
　　1916 a The relation of age to fertility in the rat. Anat. Rec., vol. 11.
　　1918 Studies on inbreeding. I. The effects of inbreeding on the growth and variability in the body weight of the albino rat. Jour. Exper. Zoöl., vol. 26.
　　1918 a Studies on inbreeding. II. The effects of inbreeding on the fertility and on the constitutional vigor of the albino rat. Jour. Exper. Zoöl., vol. 26.
　　1918 b Studies on inbreeding. III. The effects of inbreeding, with selection, on the sex ratio of the albino rat. Jour. Exper. Zoöl., vol. 27.

KING, HELEN DEAN, AND STOTSENBURG, J. M. 1915 On the normal sex ratio and the size of the litter in the albino rat (Mus norvegicus albinus). Anat. Rec., vol. 9.

KIRKHAM, W. B., AND BURR, H. S. 1913 The breeding habits, maturation of eggs and ovulation of the albino rat. Am. Jour. Anat., vol. 15.

KOLAZY, J. 1871 Über die Lebensweise von Mus rattus, varietas alba. Verhandl. zool. bot. Gesellsch., Wien.

KRAEMER, H. 1913 Über die ungünstigen Wirkungen naher Inzucht. Mittheil. der deutsch. Landwirtsch.

KUSCHAKEWITSCH, S. 1910 Die Entwicklungsgeschichte der Keimdrüsen von Rana esculenta. Festschr. zum sechzigsten Geburtstag Richard Hertwig 11. Jena.

LANTZ, D. E. 1910 Natural history of the rat. Bull. Public Health and Marine Hospital Service of the U. S. Washington, D. C.

LLOYD, R. E. 1911 The inheritance of fertility. Biometrika, vol. 8.
1912 The growth of groups in the animal kingdom. London.

LOEB, LEO 1917 The experimental production of hypotypical ovaries through underfeeding. A contribution to the analysis of sterility. Biol. Bull., vol. 33.

McCLUNG, C. E. 1902 Notes on the accessory chromosome. Anat. Anz., Bd. 20.
1902 a The accessory chromosome-sex determinant? Biol. Bull., vol. 3.

McCOLLUM, C. V. 1918 The newer knowledge of nutrition. MacMillan Co., New York.

MARSHALL, F. H. A. 1908 Fertility in Scottish sheep. Trans. Highland Agri. Soc., vol. 20.
1908 a The effects of environment and nutrition on fertility. Science Progress, vol. 2.
1910 The physiology of reproduction. London.

MILLER, NEWTON 1911 Reproduction in the brown rat (Mus norvegicus). Amer. Nat., vol. 45.

MINOT, C. S. 1891 Senescence and rejuvenation. 1. On the weight of guinea-pigs. Jour. Phys., vol. 12.

MITCHELL, A. 1865 On' the influence which consanguinity in the parentage exercises on the offspring. Edinburgh Med. Jour., vol. 10.

MITCHELL, C. W. 1913 Sex-determination in Asplanchna amphora. Jour. Exper. Zoöl., vol. 15.

MOENKHAUS, W. J. 1911 The effects of inbreeding and selection on the fertility, vigor and sex-ratio of Drosophila ampelophila. Jour. Morph., vol. 22.

MONTGOMERY, T. H. 1908 The sex ratio and cocooning habits of an aranead and the genesis of the sex ratio. Jour. Exper. Zoöl., vol. 5.
1911 The cellular basis of the determination of sex. Internat. Clinics, vol. 1.

MORGAN, T. H. 1904 Self-fertilization induced by artificial means. Jour. Exper. Zoöl., vol. 1.
1905 Some further experiments on self-fertilization in Ciona. Biol. Bull., vol. 8.
1911 An alternation of the sex-ratio induced by hybridization. Proc. Soc. Exper. Biol. and Med., vol. 8.
1914 Heredity and sex. Second edition. New York.
1914 a Two sex-linked lethal factors in Drosophila and their influence on the sex ratio. Jour. Exper. Zoöl., vol. 17.

MORGAN, T. H., PAYNE. F., AND BROWNE, E. N. 1910 A method to test the hypothesis of selective fertilization. Biol. Bull., vol. 18.

MORGAN, T. H., STURTEVANT, A. H., ET AL. 1915 The mechanism of Mendelian Inheritance. New York.

NEWCOMB, S. 1904 A statistical inquiry into the probability of causes of the production of sex in human offspring. Carnegie Institution, Washington, D. C.

NICHOLS, J. B. 1907 The numerical proportions of the sexes at birth. Mem. Amer. Anthropological Assoc., vol. 1.

OSBORNE, T. B., AND MENDEL, L. B. 1914 The suppression of growth and the capacity to grow. Jour. Biol. Chem., vol. 18.
1915 The resumption of growth after long continued failure to grow. Jour. Biol. Chem., vol. 23.
1916 Acceleration of growth after retardation. Amer. Jour. Phys., vol. 40.

OSBORNE, T. B., MENDEL, L. B., AND FERRY, E. L. 1917 The effects of retardation of growth upon the breeding period and duration of life of rats. Science, vol. 45.

PAPANICOLAOU, G. 1915 Sex determination and sex control in guinea-pigs. Science, vol. 41.

PEARL, R. 1912 Mendelian inheritance of fecundity in the domestic fowl. Amer. Nat., vol. 46.
1912 a The mode of inheritance of fecundity in the domestic fowl. Jour. Exper. Zool., vol. 13.
1913 A contribution towards an analysis of the problem of inbreeding. Amer. Nat., vol. 47.
1915 Modes of research in genetics. MacMillan & Co., New York.
1917 The selection problem. Amer. Nat., vol. 51.
1917 a Studies on the physiology of reproduction in the domestic fowl. XVII. The influence of age upon reproductive ability, with a description of a new reproductive index. Genetics, vol. 2.

PEARL, R., AND PARSLEY, H. M. 1913 Data on sex determination in cattle. Biol. Bull., vol. 24, 1913.

PEARL, R., AND PEARL, M. D. 1908 On the relation of race crossing to sex ratio. Biol. Bull., vol. 15.

PEARL, R., AND SALAMAN, R. N. 1913 The relative time of fertilization of the ovum and the sex ratio amongst Jews. Amer. Anthropologist, vol. 15, 1913.

PEARL, R., AND SURFACE, F. M. 1909 Data on the inheritance of fecundity obtained from the records of egg production in the daughters of '200-egg' hens. Me. Agri. Exper. Sta. Bull. no. 166.

PEARSON, K., LEE, A., AND BRAMLEY-MOORE. 1899 Mathematical contributions to the theory of evolution. VI. Genetic (reproductive) selection: Inheritance of fertility in man, and of fecundity in thoroughbred racehorses. Phil. Trans. Royal Soc., vol. 192.

PEARSON, K., SCHUSTER, E. H. J., AND WELDON, W. F. R. 1903 Assortive mating in man. Biometrika, vol. 2.

PFLÜGER, E. 1881 Einige Beobachtungen zur Frage über die das Geschlechtsbestimmenden Ursachen. Arch. ges. Physiol., Bd. 26.

PHILLIPS, J. C. 1914 A further study of size-inheritance in ducks, with observations on the sex ratio of hybrid birds. Jour. Exper. Zool., vol. 16.

PIKE, F. H. 1907 A critical and statistical study of the determination of sex, particularly in human offspring. Amer. Nat., vol. 41.

POPENOE, P. 1917 An experiment in long-continued inbreeding. Jour. Heredity, vol. 8.

POWYS, A. O. 1905 Data for the problem of evolution in man. On fertility duration of life and reproductive selection. Biometrika, vol. 4.

PUNNETT, R. C. 1903 On nutrition and sex-determination in man. Proc. Cambridge Phil. Soc., vol. 12.

PUNNETT, R. C., AND BAILEY, P. G. 1914 On inheritance of weight in poultry. Jour. Genetics, vol. 4.

QUAKENBUSH, L. S. 1910 Unisexual broods of Drosophila. Science, vol. 32.

RAUBER, A 1900 Der Überschuss an Knabengeburten und seine biologische Bedeutung. Leipzig.

RAWLS, E. 1913 Sex ratios in Drosophila ampelophila. Jour. Exper. Zoöl., vol. 24.

RIDDLE, O. 1914 The determination of sex and its experimental control. Bull. Amer. Acad. Med., vol. 15.

1916 Sex control and known correlations in pigeons. Amer. Nat., vol. 50.

1917 The control of the sex ratio. Jour. Washington Acad. Sci., vol. 7.

1917 a The theory of sex as stated in terms of results of studies on pigeons. Science, vol. 46.

RITZEMA-BOS, J. 1893 Onderzoekingen aangaande de gevolgen van de teelt in Bloedverwantschap. Handelingen van het vierde nederlandsche natureren geneeskundig Congres te Houden te Groningen.

1894 Untersuchungen über die Folge der Zucht in engster Blutverwandtschaft. Biol. Centralbl., Bd. 14.

ROMMELL, G. M., AND PHILLIPS, E. F. 1906 Inheritance in the female line of size of litter in Poland China sows. Proc. Amer. Phil. Soc., vol. 49.

SCHLECHTER, J. 1884 Über die Ursachen welche das Geschlecht bestimmen. Biol. Centralbl., Bd. 4.

SCHULTZE, O. 1902 Was lehren uns Beobachtung und Experiment über die Ursachen männlicher und weiblichen Geschlechtsbildung bei Tieren und Pflanzen? Sitzungsber. phys.-med. Gesellsch. zu Würzburg.

1903 Zur Frage von den geschlechtsbildenden Ursachen. Arch. mikr. Anat., Bd. 43.

SHAMEL, A. D. 1905 The effects of inbreeding in plants. Yearbook U. S. Dept. Agri.

SHULL, A. FRANKLIN 1910 Studies in the life cycle of Hydatina senta. 1. Artificial control of the transition from the parthenogenetic to the sexual method of reproduction. Jour. Exper. Zoöl., vol. 8.

1912 Studies in the life cycle of Hydatina senta. II. Internal factors influencing the proportion of male producers. Jour. Exper. Zoöl., vol. 12.

SHULL, A. FRANKLIN 1913 Inheritance in Hydatina senta. 1. Viability of the resting eggs and the sex ratio. Jour. Exper. Zoöl., vol. 15.

SHULL, A. FRANKLIN, AND LADOFF, S. 1916 Factors affecting male-production in Hydatina. Jour. Exper. Zoöl., vol. 21.

SHULL, G. H. 1910 Hybridizaton methods in corn breeding. Amer. Breeders Magazine, vol. 1.

SLONAKER, J. R. 1912 The effects of a strictly vegetable diet on the spontaneous activity, the rate of growth and the longevity of the albino rat. Pub. Leland Stanford Jr. Univ.

1912 a The normal activity of the albino rat from birth to natural death, its rate of growth and the duration of life. Jour. Animal Behavior, vol. 2.

SLONAKER, J. R., AND CARD, T. A. 1918 The effect of omnivorous and vegetarian diets on reproduction in the albino rat. Science, vol. 47.

SMITH, G., AND THOMAS, MRS. HAIG 1913 On sterile and hybrid pheasants. Jour. Genetics., vol. 3.

STEVENS, N. M. 1905 Studies in spermatogenesis with special reference to the 'accessory chromosome.' Pub. Carnegie Institution, Washington, D. C.

1911 Preliminary note on heterochromosomes in the guinea-pig. Biol. Bull., vol. 20.

1911 a Heterochromosomes in the guinea-pig. Biol. Bull., vol. 21.

STEWART, C. A. 1916 Growth of the body and of the various organs of young albino rats after inanition for various periods. Biol. Bull., vol. 31.

STOTSENBURG, J. M. 1915 On the growth of the fetus of the albino rat from the thirteenth to the twenty-second day of gestation. Anat. Rec., vol. 9.

STOUT, A. B. 1916 Self- and cross-pollinations in Cichorium intybus with reference to sterility. Mem. New York Bot. Garden, vol. 6.

SUMNER, F. B. 1909 Some effects of external conditions upon the white mouse. Jour. Exper. Zool., vol. 7.

1915 Some studies of environmental influence, heredity, correlation and growth in the white mouse. Jour. Exper. Zool., vol. 18.

THURY, M. 1864 Über das Gesetz der Erzeugung der Geschlechter bei den Pflanzen, den Thiere und den Menschen. Leipzig.

UTSURIKAWA, N. 1917 Temperamental differences between outbred and inbred strains of the albino rat. Jour. Animal Behavior, vol. 7.

WALTON, L. B. 1915 Variability and amphimixis. Amer. Nat., vol. 49.

WARREN, D. C. 1918 The effect of selection upon the sex-ratio in Drosophila ampelophila. Biol. Bull., vol. 34.

WATSON, C. 1906 Observations on diet. The influence of diet on growth and nutrition. Jour. Phys., vol. 34.

WATSON, T. B. 1905 The effects of bearing young upon the body weight and the weight of the central nervous system of the female albino rat. Jour. Comp. Neur., vol. 15.

WENTWORTH, E. N. 1913 The segregation of fecundity factors in Drosophila. Jour. Genetics, vol. 3.

WENTWORTH, E. N., AND AUBEL, C. E. 1916 Inheritance of fertility in swine. Jour. Agri. Research, vol. 5.

WHITE, F. W. 1914 Variation in the sex ratio of Mus rattus associated with an unusual mortality of adult females. Proc. Royal Soc., vol. 87.

WHITNEY, D. D. 1914 The influence of food in controlling sex in Hydatina senta. Jour. Exper. Zool., vol. 17.

1914 a The production of males and females controlled by food conditions in Hydatina senta. Science, vol. 39.

1916 The control of sex by food in five species of Rotifers. Jour. Exper. Zool., vol. 20.

1917 The relative influence of food and oxygen in controlling sex in Rotifers. Jour. Exper. Zool., vol. 24.

WILCKENS, M. 1886 Untersuchungen über das Geschlechtsverhältniss und die Ursachen der Geschlechtsbildung bei Haustieren. Biol. Centralbl., Bd. 5.

1886 a Untersuchungen über das Geschlechtsverhältniss und die Ursachen der Geschlechtsbildung bei Haustieren. Landwirthschaftliche Jahrb., Bd. 15.

WILSON, E. B. 1905 Studies on chromosomes. II. The paired microchromosomes, idiochromosomes and heterotropic chromosomes in Hemiptera. Jour. Exper. Zoöl., vol. 2.

1910 Selective fertilization and the relation of the chromosomes to sex-production. Science, vol. 32.

WILSON, JAMES 1912 The principles of stock-breeding. London.

WINIWARTER, H. VON, 1912 Etudes sur la spermatogenese humaine. Arch. Biol., Bd. 27.

WODSEDALÉK, J. E. 1913 Spermatogenesis of the pig with special reference to the accessory chromosomes. Biol. Bull., vol. 25.

WOODRUFF, L. L. 1911 Two thousand generations of paramecium. Arch. Protistenkunde, Bd. 21.

WRIEDT, C. 1916 What they say about inbreeding in Europe. Jour. Heredity, vol. 7.

YERKES, A. W. 1916 Comparison of the behavior of stock and inbred albino rats. Jour. Animal Behavior, vol. 6.

YERKES, R. M. 1913 The heredity of savageness and wildness in rats. Jour. Animal Behavior, vol. 3.

ZELENY, C., AND FAUST, E. C. 1915 Size dimorphism in the spermatozoa from single testes. Jour. Exper. Zoöl., vol. 18.

www.ingramcontent.com/pod-product-compliance
Lightning Source LLC
Chambersburg PA
CBHW080010210526
45170CB00015B/1965